4th International Conference on Advanced Composites and Materials Technologies for Arduous Applications (ACMTAA 2015)

IOP Conference Series: Materials Science and Engineering
Volume 153

Wrexham, United Kingdom
5 – 6 November 2015

Editor:

A.J. McMillan

ISBN: 978-1-5108-3176-6
ISSN: 1757-8981

Printed from e-media with permission by:

Curran Associates, Inc.
57 Morehouse Lane
Red Hook, NY 12571

Some format issues inherent in the e-media version may also appear in this print version.

Printed by Curran Associates, Inc. (2016)

For permission requests, please contact the Institute of Physics
at the address below.

Institute of Physics
Dirac House, Temple Back
Bristol BS1 6BE UK

Phone: 44 1 17 929 7481
Fax: 44 1 17 920 0979

techtracking@iop.org

Additional copies of this publication are available from:

Curran Associates, Inc.
57 Morehouse Lane
Red Hook, NY 12571 USA
Phone: 845-758-0400
Fax: 845-758-2633
Email: curran@proceedings.com
Web: www.proceedings.com

Table of contents

Volume 153

4th International Conference on Advanced Composites and Materials Technologies for Arduous Applications (ACMTAA)

5–6 November 2015, Wrexham, UK

Accepted papers received: 18 October 2016
Published online: 28 October 2016

Preface

Papers

IOP Conf. Series: Materials Science and Engineering **153** (2016) 011001 doi:10.1088/1757-899X/153/1/011001

Editorial for the proceedings of the 4th ACMTAA

AJ McMillan

Wrexham Glyndwr University, Mold Road, Wrexham, LL11 2AW

a.mcmillan@glyndwr.ac.uk

An overview of the conference

The 4th International Conference of Advanced Composites and Materials Technologies for Arduous Applications was held 5-6 November 2015, in Wrexham. This was the fourth in a series of conferences organised in collaboration between Wrexham Glyndwr University and Bauman Moscow State Technical University.

Attendees at the 4th International Conference of Advanced Composites and Materials Technologies for Arduous Applications, 5-6 November 2015, Wrexham, UK.

This was a small conference, and its primary purpose is to provide an opportunity for young researchers at both Wrexham Glyndwr University and Bauman Moscow State Technical University to present their work at an international forum, and publish it in a journal proceedings. Of course, this was an open conference, and well attended by researchers from many different organisations, from industry and academia, and from many different countries.

As an additional feature of the conference, we held a one day workshop on writing scientific journal and conference papers.

A secondary aim for the conference was to further build the collaborative relationships, not only between staff at the two institutions, but also with the many other delegates. It was unfortunate that at the time of the conference, the political situation in the Crimea was such that delegates from Ukraine were unable to obtain visas, and were sorely missed.

Finally I would like to thank all the conference participants and members of the International Scientific and Local Organising Committees, and particularly Haydn Hughes and Paul Barton for their support with the venue and catering logistics.

Editorial Team

No conference proceedings would ever be possible without the work carried out behind the scenes to manage the peer review process and the copyediting of the manuscripts. Acknowledgement and thanks are due to

Andrey Falaleev, Peter the Great St. Petersburg Polytechnic University, Russia
Martyn Jones, Wrexham Glyndwr University, UK
Rhys Jones, Monash University, Australia
Nataliia Luhyna, Wrexham Glyndwr University, UK
Galina Malysheva, Bauman Moscow State Technical University, Russia
Anna Nikolaeva, Wrexham Glyndwr University, UK
Ardeshir Osanlou, Wrexham Glyndwr University, UK
Sergey Reznik, Bauman Moscow State Technical University, Russia
Norman Swindells, Ferroday Ltd, UK
Vladimir Tarasov, Bauman Moscow State Technical University, Russia
Olga Volchek, Bauman Moscow State Technical University, Russia

Alison McMillan
School of Applied Sciences, Computing and Engineering,
Wrexham Glyndwr University

Peer review statement

All papers published in this volume of *IOP Conference Series: Materials Science and Engineering* have been peer reviewed through processes administered by the proceedings Editors. Reviews were conducted by expert referees to the professional and scientific standards expected of a proceedings journal published by IOP Publishing.

IOP Conf. Series: Materials Science and Engineering **153** (2016) 012001 doi:10.1088/1757-899X/153/1/012001

Material science problems of building space antennas with a transformable reflector 100 m in diameter

SV Reznik, PV Prosuntsov, KV Mikhailovsky, IR Shafikova

Department of Rocket and Space Composite Structures, Bauman MSTU, 5/1, 2nd Baumanskaya str., Moscow, 105005, Russia

E-mail: sergeyreznik@bmstu.ru

Abstract. This paper focuses on the development of a space antenna transformable reflector 100 m in diameter. Accuracy requirements to the materials characterization are based on the thermal state analysis. The paper contains methods and results of thermal physical and optical characterization of the main construction materials to use in the transformable reflector.

1. Introduction

Space communication development, natural resources exploration, and space observations require deployable space reflectors to be developed. According to AstroMesh estimates, the existing technology enables mesh reflectors with a diameter of up to 50 m within the next decade.

Low linear density, 0.3 kg/m^2 in the best cases, is achieved by using radio-reflecting metal meshes made from tungsten, molybdenum or nichrome wires of 15-50 microns in diameter. The reflecting surface is geometrically stable, owing to bracing wires, hinges, props, struts, and power cables. The load-bearing elements are made in the form of straight or curved rods from polymer matrix composite materials, primarily carbon fibre-reinforced plastics.

The greatest challenge when designing space antenna reflectors is to ensure high dimensional stability under varying temperatures and temperature differences in the orbit. The permissible variation in shape and size should not exceed $\Lambda/16$-$\Lambda/50$, where Λ is the radiation wavelength. In the L frequency band, where transformable reflectors are mostly used, these variations constitute 4-14 mm. A transformable reflector should be compactly stowed under the rocket booster fairing, whose diameter normally does not exceed 4 m.

At this stage of research, a theoretical and technological foundation must be created in the field of designing and manufacturing large transformable antenna reflectors up to 100 m in diameter, with a high-precision reflecting surface for long-life spacecraft.

2. The current state of research in the field of transformable reflectors of space mirror antennas

Large transformable antenna reflectors are currently being developed in Russia, the USA, Europe, China and Japan.

The Russian Joint Stock Company (JSC) Academician M.F. Reshetnev Information Satellite Systems installs umbrella-type mesh reflectors onto the Luch-5 geostationary relay satellites [1, 2].

For example, the Luch 5B has two antennas with umbrella-type mesh reflectors 4.2 m in diameter that can focus on low-orbit spacecraft, capture and follow them along the flight path. One of these antennas operates in Ku-band, the other in S-band.

The JSC S.P. Korolev Rocket and Space Corporation "Energia" has developed several distinctive designs of large transformable reflectors for space antennas [3, 4]. A 6.4 m by 5.2 m model of one of the reflectors was produced by JSC NPO "EGS" (a joint enterprise of "Energia" and the Georgian Polytechnical Institute, Georgia) and tested in the Mir orbital station in 1999 (figure 1).

Figure 1. Mesh reflector prototype during orbital testing at the Mir space station (JSC NPO "EGS", Russia – Georgia).

The JSC Special Design Bureau of Moscow Power Engineering Institute (OKB MEI) has developed transformable reflectors for radar space antennas (figure 2). TKCA-6K antennas with 6 m by 2.8 m reflectors were successfully tested in orbit in 1985, as part of synthetic aperture radar systems onboard the Kosmos 1689 spacecraft, and in 1996, onboard the Priroda module of the Mir space station. The improved antennas are to be used as part of the Kondor remote probing satellites by the JSC Military Industrial Corporation Scientific-Production Association of Machine building (MIC SPA) "Mashinostroyenia" [5]. Kosmos-2487, the first satellite of this type deployed in 2013, had a 6 m by 7 m antenna reflector. The hollow rods of the load-bearing frame were not made from metal, as before, but from carbon fibre-reinforced plastics. The overall reflector mass including hold-up mechanisms and all structural elements did not exceed 50 kg. The multi-beam feed system enabled both the beam mode and SCANSAR mode.

The USA-based Harris Corporation equipped the TerreStar, TDRS, ACeS, SDARS satellites with the umbrella-type transformable reflectors, with diameter up to 18 m (figure 3) [6]. The AstroMesh, Northrop Grumman branch, developed the largest transformable hoop type (membrane-cable) reflectors for the Thuraya-3, MBCO, Inmarsat, and GlobalStar satellites. AM-Lite, AM, AM-1, AM-2 type reflectors have apertures of 3-8 m, 12 m, 6-25 m and 18-50 m respectively (figure 4) [7, 8].

Specialists in a number of companies in Europe and Japan [9-12] are also working on the development of large transformable reflector.

3. Reflector structural design

Contemporary space antenna reflectors comprise load-bearing rod elements to deploy the reflector into the operating configuration. They form the frame for the metal mesh comprising the radio-reflecting surface. Another essential element is polymer ropes, adjusting and supporting the mesh surface and metallic tips jointing the ropes to the frame.

Figure 2. Mesh reflector of the space antenna during ground tests (JSC "OKB MEI", Russia).

Figure 3. Deployed TerreStar-1 18m reflector at Harris facility.

Designing transformable reflectors for space antennas is a complex interdisciplinary problem, where the initial stages are concerned with the thermal analysis and materials science issues. One of the design layouts for a reflector 100 m in diameter is analyzed below (figure 5). The load-bearing structure comprises carbon composite spokes that unfurl in the orbit forming the supporting ring for the mesh deployment. The parabolic shape is enabled by the tensioning ropes.

IOP Conf. Series: Materials Science and Engineering **153** (2016) 012001 doi:10.1088/1757-899X/153/1/012001

4. Reflector temperature state simulation

The flight of a spacecraft with a reflector, on a geostationary orbit, at the March equinox moment was considered. It was assumed that the rotation axis of the parabolic reflector shell is directed toward the Earth and its transverse axis coincides with the spacecraft velocity vector.

Figure 4. Deployed Reflector at AstroMech Corporation.

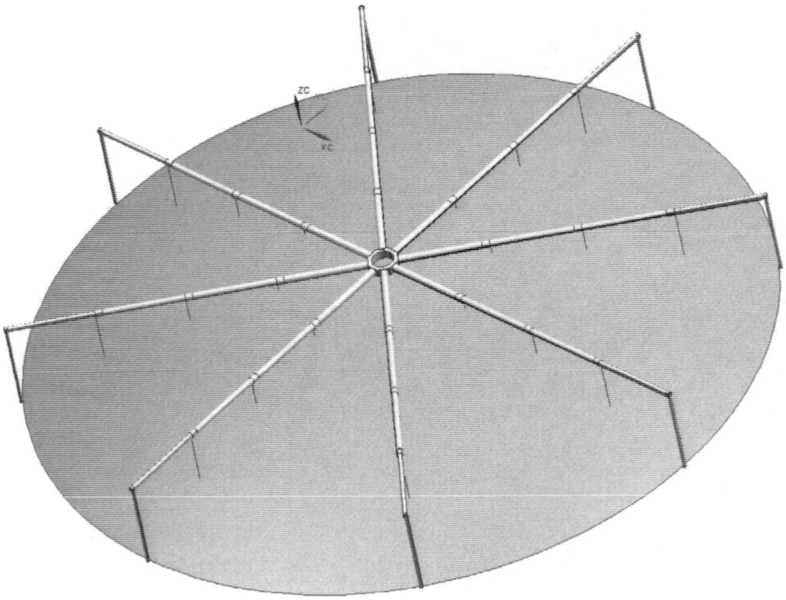

Figure 5. Reflector design layout.

The mesh was assumed to reflect, absorb and transmit the incident direct and Earth-reflected solar radiation. The thermal physical anisotropy of the carbon composite frame was taken into account.

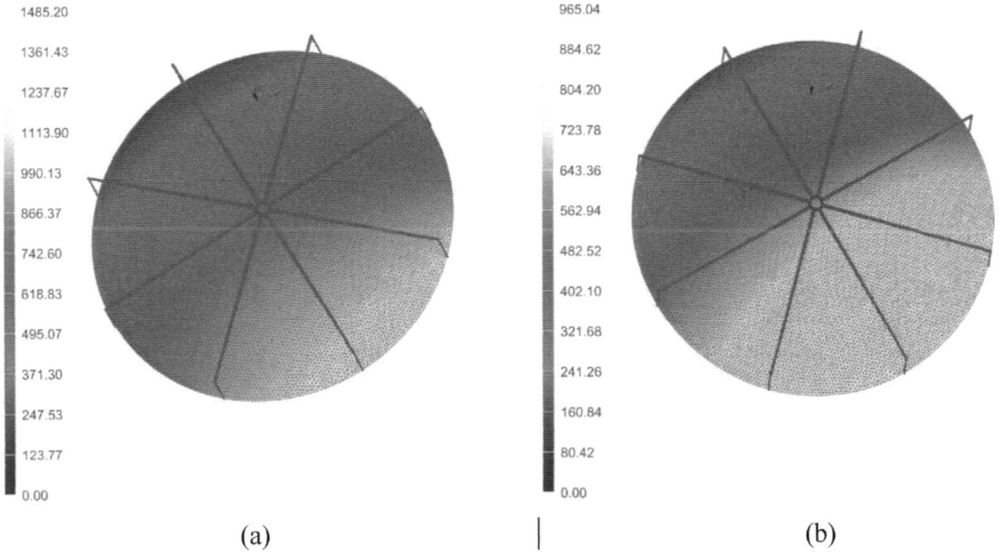

Figure 6. (a) Incident and (b) transmitted solar radiation flux when the angular position of the reflector in orbit, relative to the Earth-Sun line, is 120°, W/m^2

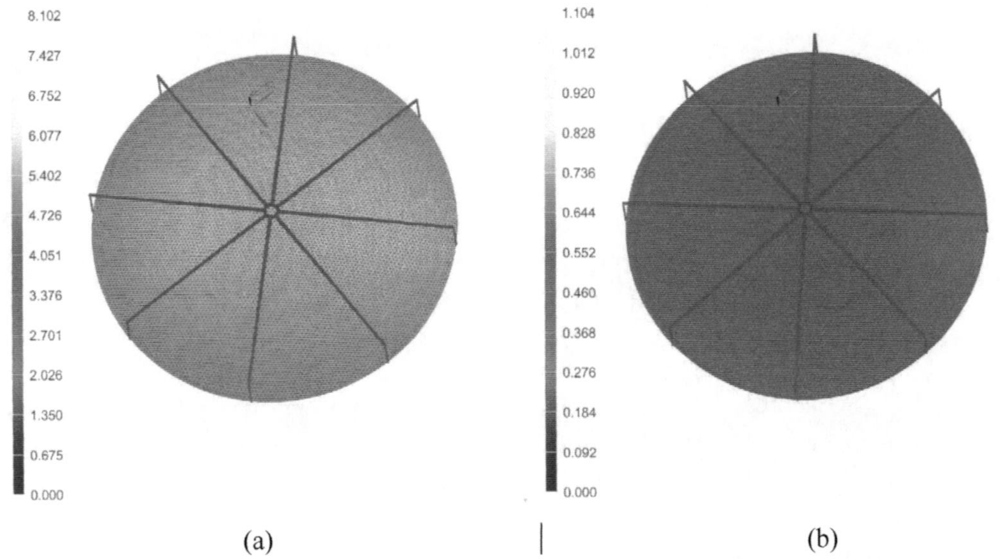

Figure 7. (a) Intrinsic and (b) Earth-reflected solar radiation flux when the angular position of the reflector in orbit, relative to the Earth-Sun line, is 120°, W/m^2

A finite-element model of the reflector was built, using approximately 60,000 3D tetragonal elements for composite arms, and approximately 30,000 2D elements for the grid-curtain and cables. The thermal contacts between separate structure elements were considered to be ideal. The surface of the carbon fire reinforced plastic (CFRP) was considered to be diffusely radiating, and the cables semi-transparent for downward radiation. The thermal-physical properties of CFRP were taken to be: density 1550 kg/m3, and specific heat 1000 J/(kg·K). The thermal conductivity across the fibres was taken to be 72 W/(m·K), and the values of thermal conductivity along the fibres was variable: 101, 72, 36, 18 and 9 W/(m K). The emissivity factor and radiation capacity for CFRP were equal to 0.7. The density of the cable material was 1000 kg/m3, specific heat 800 J/(kg·K), and thermal conductivity 1 W/(m·K). The cable emissivity was equal to 0.05, its radiation absorptivity 0.50, and the cable material transmitivity was 0.45.

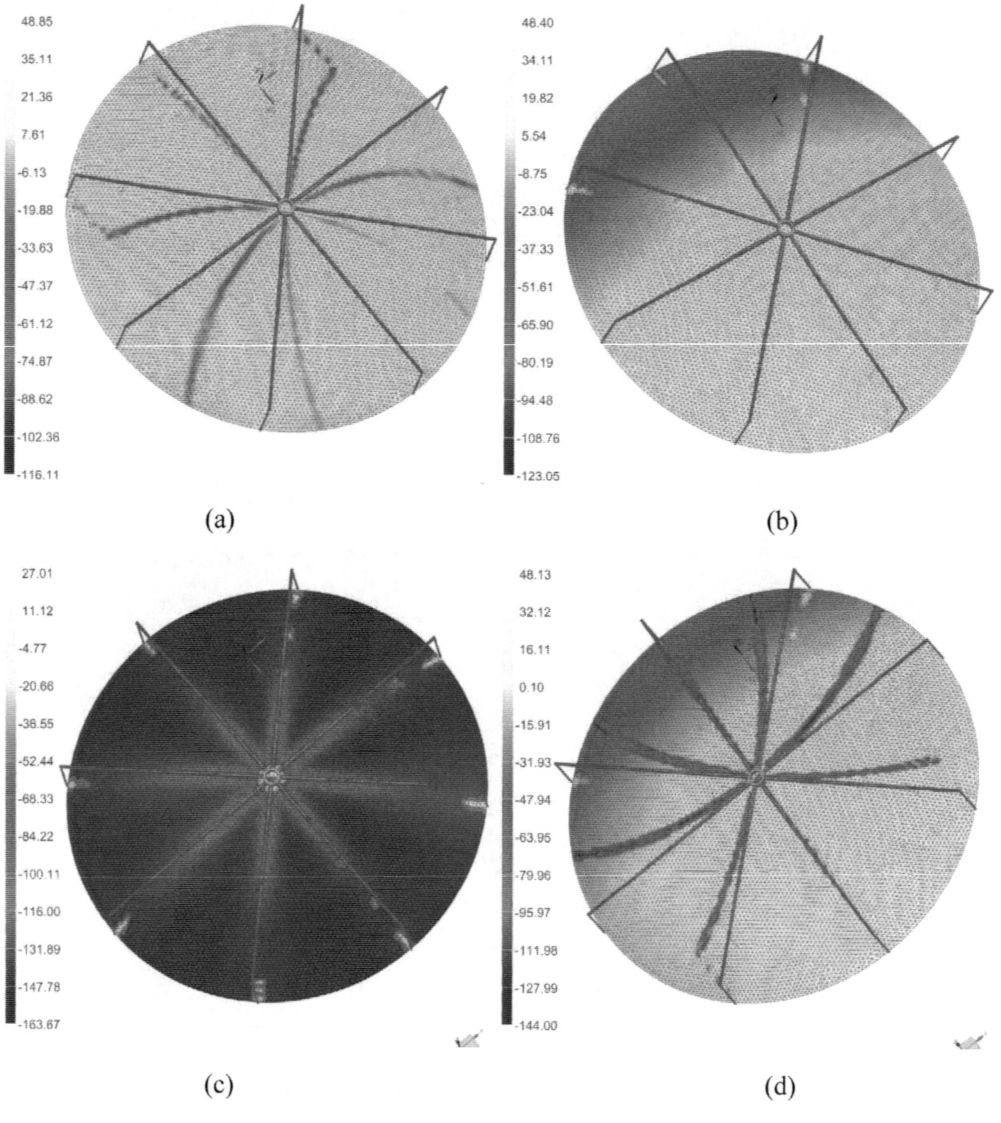

Figure 8. Reflector temperature field, °C when the angular position of the reflector in orbit, relative to the Earth-Sun line, is (a) 90°, (b) 120°, (c) 180° and (d) 300°.

The metal mesh was modelled as a partially transparent shell with the thickness 0.025 mm produced from material with density 3200 kg/m^3, specific heat 500 J/(kg·K), and thermal conductivity varying in the range from 0.45 to 5.40 W/(m·K). The metal mesh had transmission capability 0.85, and its emmisivity and radiation absorptivity were equal to 0.09.

Incident and Earth-reflected solar radiation flux and flow of the Earth thermal radiation were defined for the temperature state simulation using the Siemens PLM NX 10finite element analysis package (figures 6, 7). Thermal state modelling also performed with the Siemens PLM NX 10 software package demonstrated temperature differences in the reflector structure as high as 190 K (figure 8).

5. Results analysis

When designing a reflector, reliable initial data on thermal physical and optical materials characteristics are essential. Unfortunately, there is limited information on the thermal conductivity coefficient in the reinforcement plane for composite materials, which is due to the absence of a standardized methodology. In the case of the load-bearing structure, we should calculate the possible effect of its thermal conductivity coefficient error on its temperature field. Figure 9 shows the temperature field of the load-bearing frame at different values of the rod material thermal conductivity coefficient. It is evident that the difference in the spoke temperature values reaches 30 K, which indicates the need to increase the accuracy of the thermal conductivity estimation.

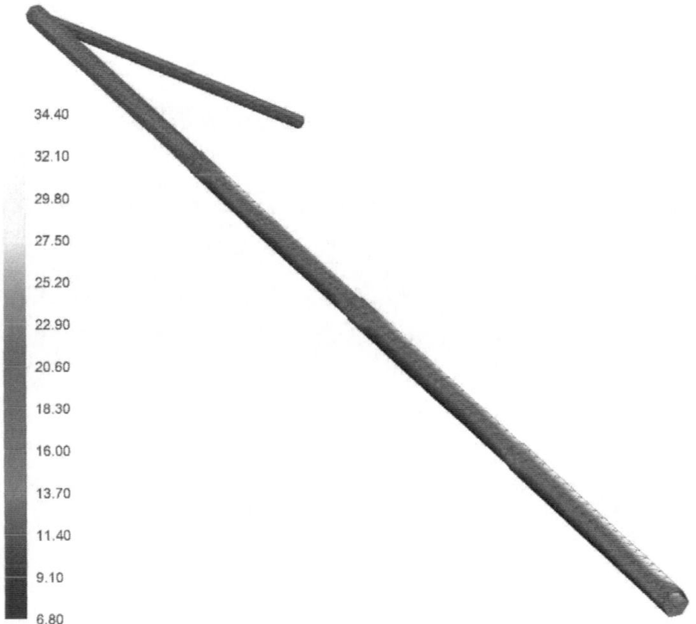

Figure 9. Temperature field, °C, of the reflector rod at longitudinal thermal conductivity 72 W/(m·K), when the angular position of the reflector in orbit, relative to the Earth-Sun line, is 120° .

In order to compensate for the deficiencies in the reference literature the authors have developed novel methods of optical characterization of metallic meshes and polymer ropes, as well as a method to determine the thermal conductivity of composite materials directly in the full-scale structural elements in the form of rods and plates [13-17].

The fact is that the traditional methods [18-21] allow defining of thermal conductivity of composites only in the normal direction $\lambda_{\perp}$. Experimental samples are round or quadrangular in shape and machined very carefully. Information about $\lambda_{\perp}$ as well as determination accuracy is not of the utmost interest when modelling thermal regimes of thin-walled space structures due to a slight temperature difference through-the-thickness that is equal to just a few degrees. The value of the longitudinal thermal conductivity (in-plane direction along the fibres) $\lambda_{\parallel}$ is much more important because it influences to the temperature difference over the surface and along the structure. A peculiarity of this method for defining λ is the possibility to use full-size structures such as hollow rods [13, 14]. At the same time, this method is suitable for cables thermal conductivity research. Experiments could be conducted with one-sided [13, 14] and two-sided [15-17] heating of samples. The temperature difference is generated by electrical heating elements and measured by thermocouples or by a thermal imager. A program for solving the nonlinear inverse heat conduction problem is used for the interpretation of experimental data [22-23].

Obtaining reliable data on the mesh thermal conductivity coefficient is a considerable challenge, because of its flexibility when measuring tension. At present, there is practically no information on the thermal conductivity coefficient of the mesh. The thermal conductivity varying in the range 0.45 to 5.40 W/(m·K) indicates that it has only a very slight effect on the reflector thermal state. This must be a result of the fact that with the reflector being large and the mesh being thin, each point of the mesh is close to radiative equilibrium, when the conductive fluxes are significantly smaller than the radiative ones.

At the same time, the variability of the mesh optical characteristics indicates its significant effect on the reflector thermal state (figure 10).

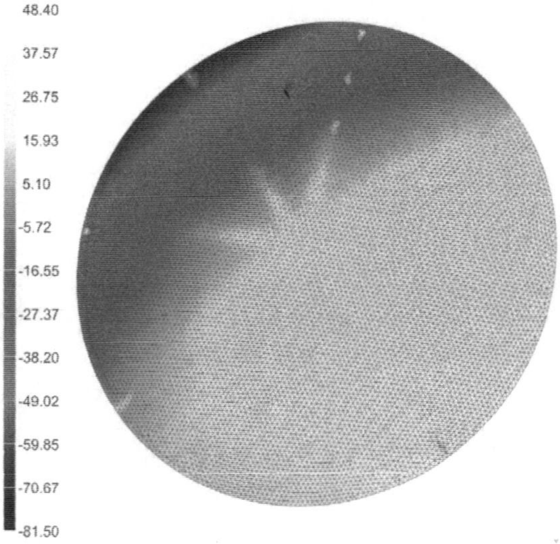

(a)

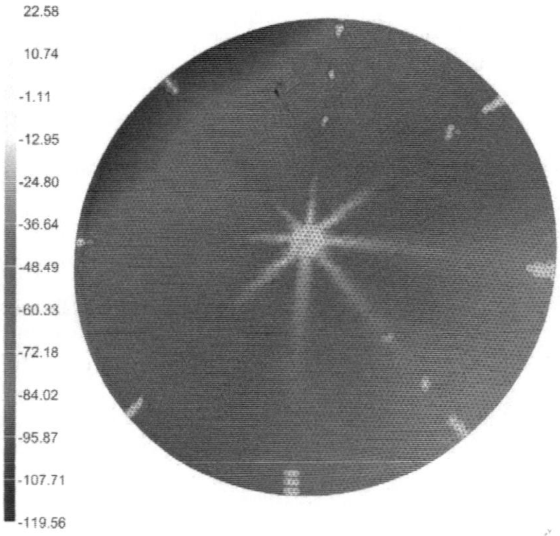

(b)

Figure 10. Mesh temperature field, °C, at different values of optical characteristics at time 28800 sec. Thermal conductivity – 3.6 W/(m·K)

a) Emissivity and absorptivity – 0.7, transmittivity – 0.85
b) Emissivity – 0.135, absorptivity –0.09, transmittivity – 0.85

6. Conclusions

Several transformable reflectors layouts for large diameter space antennas have been considered. The common feature for all the layouts in question is the combination of the load-bearing elements from carbon fibre-reinforced plastics, a metallic radio reflecting mesh and polymer ropes for the surface shape adjustment

The thermal state of a 100 m diameter reflector was modelled for the geostationary orbit, which demonstrated a temperature difference of 190 K in the reflector structure.

The composite material thermal conductivity coefficient was demonstrated to have a significant influence on its temperature state.

At the same time, the mesh thermal conductivity coefficient influence is insignificant, with the data on its optical characteristics being of much greater importance.

7. Acknowledgment

Some results of this work were obtained in the framework of the Agreement on grant No. 14.577.21.0129 with the Ministry of education and science of the Russian Federation, unique identifier for applied scientific research (project) RFMEFI57714X0129.

References

[1] Bessmertnaya JS, Kyzovnikov AV, Vygonskiy YG, and Viryasov MM 2014 Система ретрансляции «Луч» *Решетневские чтения: Матер. XVIII Междунар. науч. конф., посвящ. 90-летию со дня рождения генер. конструктора ракет.-космич. систем акад. М.Ф. Решетнева* 11–14 нояб. 2014, г. Красноярск, Россия **1** ["Retransmission system "Luch" *Proc. 18th Int. Sci. Conf. Dedicated to Acad. M.F. Reshetnev,* Krasnoyarsk, Russia **1**, 173-174 (in Russian)]

[2] Retransmission Space System "Lych", http://www.iss-reshetnev.ru/projects

[3] Gottero M, Sacchi E, Lorenzo G, Reznik SV and Kalinin DYu, 2005, *Proc. 35th Int. Conf. on Environmental Systems,* Roma, Italy

[4] Reznik SV, Kalinin DYu and Denisova LV, 2010, *Proc. 32nd ESA Antenna Workshop on Antennas for Space Applications. From technologies to architectures*, Noordwijk, The Netherlands, 8

[5] Afanas'ev I, 1999, *Astronautics News* (Novosti Kosmonavtiki) **10** (201)

[6] Semler D, Tulintseff A, Sorrell R and Marshburn J, 2010, *Proc. 28th AIAA International Communications Satellite Systems Conference -ICSSC-2010*, Anaheim, California, USA, AIAA-2010-8855

[7] Thomson MW, 2002, *Proc. 20th AIAA International Communications Satellite Systems Conference ICSSC-2002*, Montreal, Canada, AIAA-2002-2032

[8] Marks G, Lillie C and Kuehn S, 2011 *Proc. 33rd ESA Antenna Workshop on Challenges for Space Antenna Systems. Preparing for the Future*, Noordwijk, The Netherlands

[9] Ihle A, Breunig E, Datashvili L, Migliorelli M, Scialino L, van't Klosters K and Santiago-Prowald J, 2012, *Proc. Workshop on Large Deployable Antennas,* Noordwijk, The Netherlands

[10] Sinn T, Proetti Zola P, Pfeiffer EK, Datashvily L and Lori M, 2015, *Proc. 36th ESA Antenna Workshop on Antennas and RF Systems for Space Science*, Noordwijk, The Netherlands

[11] Meguro A, Harada S and Ueba M, 2005, *Proc. 56th International Astronautical Congress*, Fukuoka, Japan, IAF-05-C2.1.B

[12] Ozawa S, Shintate K, Tsujihata A, 2012, *Proc. Workshop on Large Deployable Antennas*, Noordwijk, The Netherlands

[13] Reznik SV, Prosuntsov PV, Railyan VS and Shulyakovsky AV, 2007, *Proc. Symposium on*

Inverse Problems, Design and Optimization, Miami, Florida,USA

[14] Reznik SV, Denisov OV, Prosuntsov PV, Timoshenko VP and Shulyakovskiy AV, 2012, Термовакуумные испытания полых композитных стержней для космических конструкций *Vse Materialy. Entsiklopedicheskij Spravochnik* **7**, 8-12 [Thermal vacuum tests of hollow composite rods for space structures *All the Materials. Encyclopedic Reference Book* **7**, 8-12 (in Russian)]

[15] Reznik SV, Timoshenko VP, Prosuntsov PV, Minakov DS 2013 "Моделирование и идентификация параметров теплопереноса в тросовых элементах космических конструкций. I. Обоснование условий эксперимента" *Тепловые процессы в технике* **5**(5) 235-9 ["Modeling and identification of heat transfer parameters in rope elements of space structures. I. The rationale for the experimental conditions" *Thermal Processes in Engineering* **5**(5) 235-9 (in Russian)]

[16] Reznik SV, Timoshenko VP, Prosuntsov PV, Minakov DS 2014 "Моделирование и идентификация параметров теплопереноса в тросовых элементах космических конструкций. II. Экспериментальные исследования. Определение теплопроводности тросового элемента" *Тепловые процессы в технике* **6(8)** 378-84 ["Modeling and identification of heat transfer parameters in rope elements of space structures. II. An experimental study. Determination of the thermal conductivity of the rope element" *Thermal Processes in Engineering* **6(**8) 378-84 (in Russian)]

[17] Denisov OV, Kirbai AA and Minakov DS, 2015, *Proc. Conf. Heat and Mass Transfer in the Thermal Control System of Technical and Technological Energy Equipment,* Tomsk, Russia, **23**, 01016 Doi: 10.1051/Matecconf/20152301016

[18] ASTM C 177: Standard Test Method for Steady-State Heat Flux Measurements and Thermal Transmission Properties by Means of the Guarded-Hot-Plate Apparatus http://www.astm.org/Standards/C177.htm

[19] Johansson P, Adl-Zarrabi B and Hagentoft CE, 2012, *Frontiers Architect. Res.* **1**(4) 334-40.

[20] Krupa P and Malinaric S, 2014, *Int. J. of Mathematical, Computational, Physical, Electrical and Computer Engineering*, **8**(5) 735-40

[21] Ponomarev SV, *et al*, 2008, *Теоретические и практические основы теплофизических измерений* ФИЗМАТЛИТ, Москва [*Theoretical and practical bases of thermophysical measurements* FIZMATLIT, Moscow (in Russian)]

[22] Reznik SV *et al* 1997 "Математико-алгоритмическое и программное обеспечение исследования процессов радиационно-кондуктивного теплообмена в частично прозрачных материалах" *Труды 1-го междунар. симп. Передовые термические системы и технологии,* Katsiveli, Crimea, Ukraine, Bauman MSTU Publisher House, Moscow 40-49 ["Mathematical algorithms and software for the study of processes of radiative-conductive heat exchange in partially transparent materials" *Proc. 1st Int. Symp. Advanced Thermal Engineering Systems and Technologies Part 2,* Katsiveli, Crimea, Ukraine, Bauman MSTU Publisher House, Moscow (in Russian)]

[23] Polezhaev YuV, *et al*, 2002, *Материалы и покрытия в экстремальных условиях. Взгляд в будущее: В 3 т. – Т. 1. Прогнозирование и анализ экстремальных воздействий* vol. 1, Ed. SV Reznik, Bauman MSTU Publisher House, Moscow [*Materials and Coatings in Extreme Conditions. Looking to the future. Forecasting and Analysis of Extreme Impacts* vol. 1, Ed. SV Reznik Bauman MSTU Publisher House, Moscow (in Russian)]

Optimization of the curing process of a sandwich panel

Pyi Phyo Maung, O Tatarnikov , G Malysheva

Russia, Moscow, 2nd Baumanskaya Street, 5, Bauman Moscow State Technical University

pyaephyo@mail.ru

Abstract. This study presented finite element modelling and experimental measurements of temperatures during the autoclave curing of the T-50 aircraft wing sandwich panel. This panel consists of upper and lower carbon fibre based laminates and an aluminium foil honeycomb. The finite element modelling was performed using the Femap-Nastran product. During processing, the temperature at various points on the surface of the panel was measured using the thermocouples. The finite element method simulated the thermal conditions and determined the temperatures in the different parts of the panel for a full cycle of the curing process. A comparison of the calculated and experimental data shows that their difference does not exceed 6%.

1. Introduction

Composite sandwich panel (three layered) structures are a combination of relatively thin monolithic panels (carbon fibre skins), between which the honeycomb is located. The main purpose is to provide rigidity. The outer composite layers are the basic support elements for the load.

Nowadays composite sandwich panel structures are widely used in the manufacture of aircraft and helicopter fairings; fuselage and wing panels; spars, frames and other structural parts [1, 2]. Aerospace applications for sandwich panel structures include rocket nose cones, the frontal screens of landers; shell compartments; frames of solar panels, etc.. The application of polymer composites is also increasing every year in the construction, automotive and aerospace industries [3, 4].

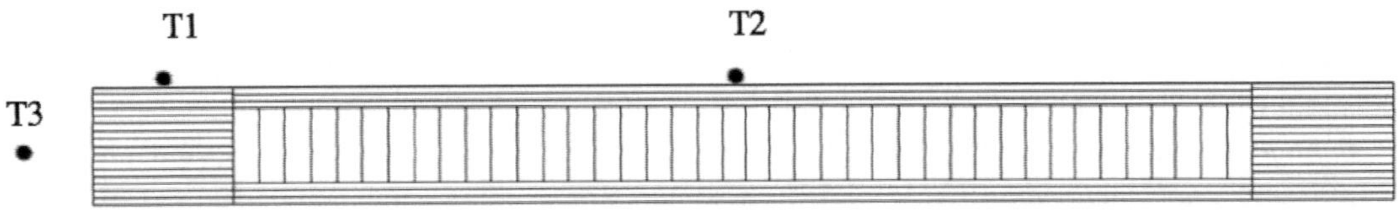

Figure 1. Schematic sandwich panel indicating thermocouples: T1, T2 and T3

The main objective of this paper is to study the composite sandwich panel wing structure from a tactical 5th generation Russian fighter aircraft. A carbon fibre based epoxy binder pre-preg was used in the manufacture of the wing panel.

A simplified sketch of the composite sandwich panel is shown in figure 1. The difference from other traditional sandwich wing panel structures is that the honeycomb panel is only placed in small portions where there is a need to provide rigidity. Other portions of the panel are composed solely of carbon fibre laminate. Although the heating temperature in autoclave for the combined structure is the same, the temperatures at the portion of the composite laminate skin with and without an included honeycomb panel are significantly different. The thermocouples are located as shown in figure 1 and they are identified as follows: the temperature on the monolithic portion (laminate) was described as T1, the temperature at sandwich panel was T2 and temperature in the autoclave was T3.

The aim of the paper is to develop a finite element model to evaluate the temperatures at various locations of the panel structure and to compare with the experimental measurements during the curing process in the autoclave.

2. Methodology

The temperatures at the different parts of the wing panel were measured experimentally during the full curing cycle in autoclave. Thermocouples of the type TMKn, manufactured by CRE-IRC, were used, which are comprised of chromel-constantan, and operate in the range of 0 °C to 800 °C.

Initially, the temperature in the panel was 20 °C, and then it was placed in an autoclave, where the heat was applied from all sides of the panel is shown in Figure 2. The heating rate was 2 °C per minute, and the cooling rate in the autoclave, but at normal atmospheric pressure, was -1 °C per minute. The first exposure was carried out at 125 °C for 60 minutes and the second was at 180 °C for 300 minutes.

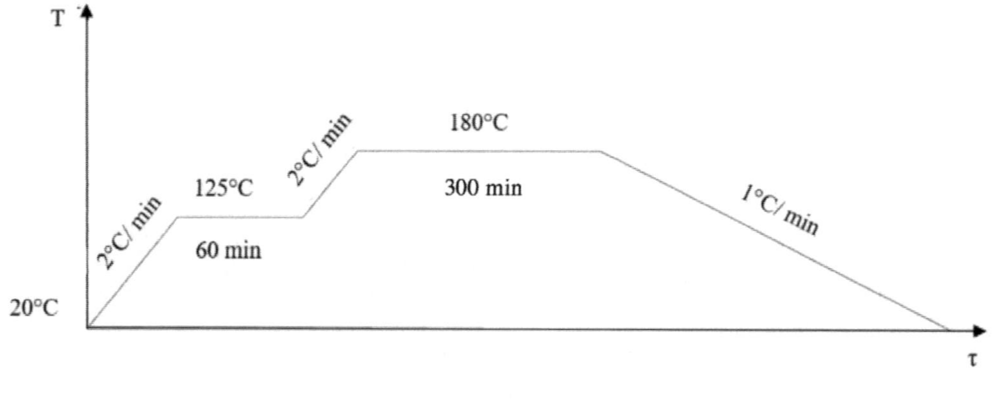

Figure 2. Sandwich panel curing in an autoclave (temperature T vs time τ)

The real temperature profile differed from the pre-set heating (figure 3), since the thermal properties of the solid laminate section and the section where with honeycomb were different. The data shows that the lowest recorded temperature values from thermocouple T1 were on the laminate panel section. The experiment determined that if the actual temperature was lower than the programmed temperature then the area will not be fully cured. The mechanical properties of the structure were thus not fully developed.

Experimental methods for the evaluation of the curing mode, depending on the geometric characteristics, are widely used in designing the composite structures [5-8]. Their main disadvantages are high cost and that the results cannot be used in other designs, even where they are very similar in structure. That is why numerical methods are widely used for modelling of the curing of polymeric composite materials [9-11].

In this study, a finite element model was built to investigate composite the sandwich panel and to simulate the temperature during the curing process by using the software package FEMAP-NASTRAN.

In FEMAP-NASTRAN the composite laminate package was simulated with three-dimensional solid elements, and the aluminium honeycomb was treated as a plate element.

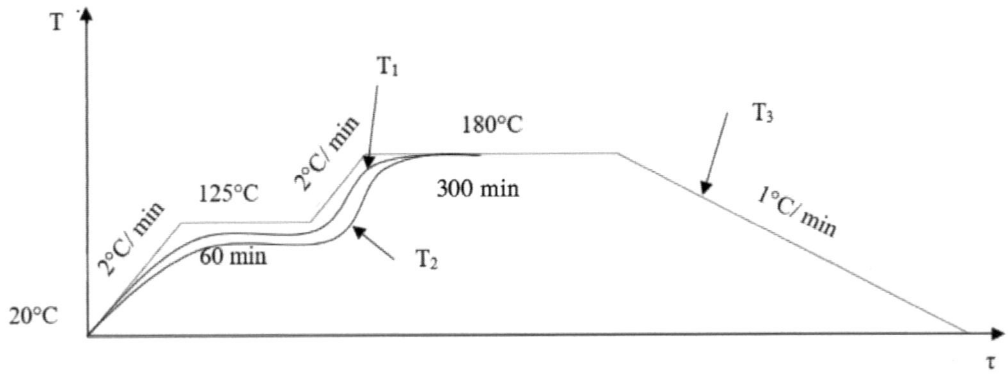

Figure 3. The actual curing sandwich panel in the autoclave (dependence of temperature T and time τ)

Table 1. The geometrical parameters of the panel and the materials used

Material skins	CFRP UWC-30K.R14535 (adhesive pre-preg brand KMKU-3m.150.R14535.45 "Porsher")
Honeycomb material	Cell (01-11 OST 1 00728) Foil AlMg2N
Skin thickness	2 mm
Honeycomb thickness	4 mm
Monolayer thickness	0.14 mm
Panel size	0.6 × 0.8 m
Plate thickness	4-6 mm

The problem was solved as a non-steady state heat conduction model. The initial conditions assumed a uniform temperature field of 20 °C across the model. The boundary conditions were specified as convective heat transfer from the air, which varied according to the specified temperature mode (figure 3). The heat transfer coefficient depends on the distance between the surface of the panel, the wall of the autoclave and the ventilation characteristics of the chamber, as well as on the location of the heating element relative to the heated panel. The geometrical parameters of the panel, its elements, and the materials used are shown in table 1.

3. Results

The model developed allowed the stimulation of the state of thermal cure and determination of the temperature values at different parts of the three-layer panel for every time step. The modelling results at temperature step 180 as shown in figure 4.

Figure 4. The modelling results in Femap Nastran at time step – 7610 sec
The temperature range is 160 °C (blue) to 170 °C (red)

The modelling results and experimental values of temperatures on the surface of the panel are shown in table 2. The results in table are related to the time taken to reach the set temperature and referred to as the time of exposure (holding time) at a given temperature. The accuracy of the modelling process was calculated based on experimental results being 100%.

Table 2. Time values determined experimentally and theoretically

№	Temperature, °C	Time, minutes		Accuracy, %
		Experimental	Theoretical	
1	180	187	198	5.6
2	251	192	202	4.9

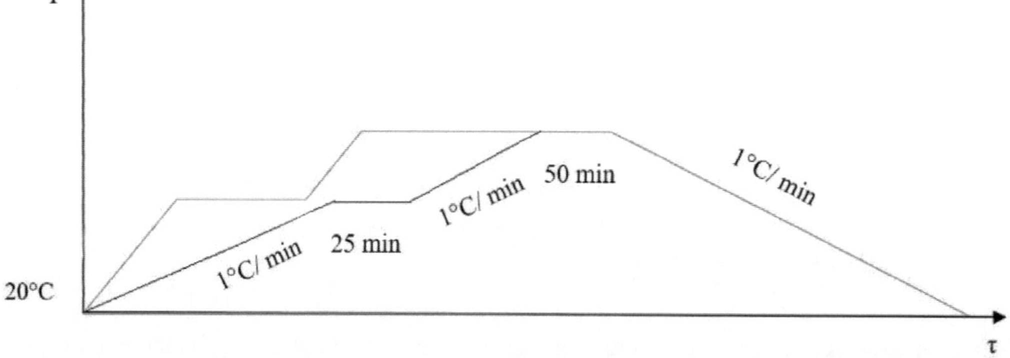

Figure 5. The proposed cure cycle for sandwich panel structure in Femap Nastran

4. Discussion

Comparison of simulation results and experimental data shows good agreement. The maximum relative difference does not exceed 6%. This gives confidence in the simulation to be able to evaluate the cure kinetics parameters of the wing sandwich panel structure. After taking account the all parameters of the curing process, the optimal cure cycle for sandwich panel structure in the autoclave is proposed (figure 5).

The modelling results using proposed cure cycle are shown in figure 6. The results show that the proposed cure cycle is optimal for sandwich panel structure in the autoclave.

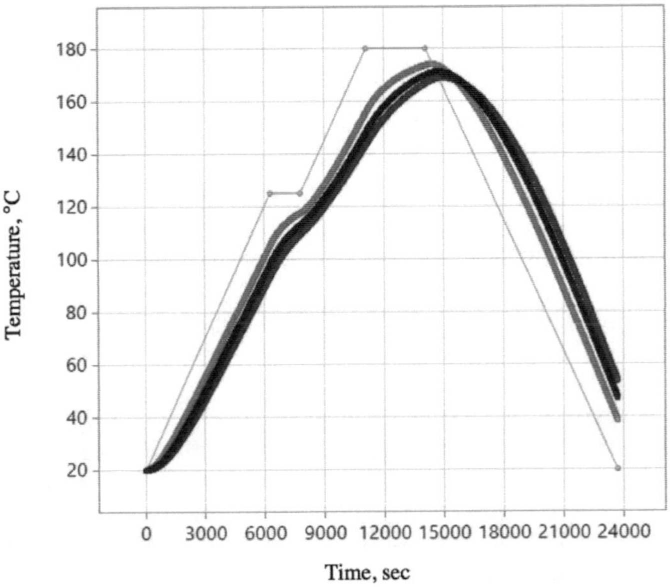

Figure 6. The modelling results by using the proposed optimal cure cycle in Femap Nastran

5. Conclusions

The study showed that the temperatures at various portions of the panels during the experiment were programmed to have different temperatures. This was due to significant difference in thermal properties of various areas of the panel.

The temperature at three points was determined experimentally, which allowed the curing time to be adjusted. Thus, the exposure time (holding time) can be optimized for the two isothermal stages of heating (125 °C and 180 °C for the given material).

The theoretical model of sandwich panels developed allows the determination of the temperatures in different parts of the panel during the curing process.

Comparison of the calculated and experimental results showed the maximum relative difference between them did not exceed than 6%. The finite element model developed can thus be used for the evaluation of temperature of sandwich panel structure in the autoclave.

References

[1] Журавлева КВ, Шевцова МА, Малышева ГВ 2013 Расчет элемента конструкции крыла самолета, изготовленного из полимерных композиционных материалов *Энциклопедия инженера-химика* **8** 27–30 [Zhuravleva K, Shevtsova M and Malysheva G The calculation of the structural element of an aircraft wing made from polymer composites *J. Encyclopedia of Chemical Engineering* (in Russian)]

[2] Волков ЕО, Разина АС, Нехороших ГЕ 2012 Разработка модели каркаса развертываемого объемного модуля на основе упругих элементов из полимерного композиционного

материала *Все материалы. Энциклопедический справочник* **8** 31–33 [Volkov E, Razina A and Nehoroshih G Development of a model framework deployed on the basis of the bulk modulus of the elastic elements of a polymeric composite material *All materials. Encyclopedic Reference* (in Russian)]

[3] Нелюб ВА, Карасева АА, Боченкова АА 2012 Конструкционные стеклопластики на основе полиэфирной матрицы *Все материалы. Энциклопедический справочник* **7** 46–49 [Nelyubov I, Karaseva A and Bochenkova A Construction GRP on basis polyester matrix *All materials. Collegiate directory* (in Russian)]

[4] Кузнецов ВМ, Нехороших ГЕ 2015 Применение углепластиков при изготовлении сосудов давления торовой формы *Все материалы. Энциклопедический справочник* **2** 32–36 [Kuznetsov V and Nehoroshih G Application of carbon fiber reinforced plastics in the manufacture of pressure vessels torus *All forms materials. Collegiate directory* (in Russian)]

[5] Беляков ЕВ, Тарасов ВА, Боярская РВ 2012 Выбор режимов формования композитных конструкций ракетно-космической техники *Известия высших учебных заведений. Машиностроение* **5** 37–43 [Belyakov E, Tarasov V and Boyarskaya P Moulding composite designs space rocket technology *News higher training institutions. Engineering* (in Russian)]

[6] Комков МА, Тарасов ВА, Зарубина ОВ 2013 Определение толщины композитного слоя сферической оболочки с учетом ширины наматываемой ленты *Известия высших учебных заведений. Машиностроение* **10** 20–29 [Komkov M, Tarasov V and Zarubina O Determination of the thickness of the composite layer spherical shell taking into account the width of the tape wound *Proceedings of the higher educational institutions. Mechanical Engineering* (in Russian)]

[7] Филимонов АС, Тарасов ВА, Комков МА, Моисеев ВА, Тимофеев МП, Герасимов НВ 2012 Экспериментальный анализ свойств перспективных теплоизоляционных материалов машиностроения, полученных методом фильтрационного осаждения *Инженерный журнал: наука и инновации* **9**(9) 38 [Filimonov A, Tarasov V, Moiseev V, Timofeev M and Gerasimov N Experimental analysis of the properties of thermal insulation materials engineering perspective, produced by filtration deposition *Engineering Journal: science and innovation* (in Russian)]

[8] Гузева ТА 2014 Методы оценки свойств связующих, применяемых при производстве изделий из полимерных композиционных материалов *Все материалы. Энциклопедический справочник* **5** 22–24 [Guzeva T Methods evaluation CTB used in the manufacture of polymer composite materials *All materials. Encyclopedic Reference* (in Russian)]

[9] Лурье СА, Миронов ЮМ, Нелюб ВА, Бородулин АС, Чудной ИВ, Буянов ИА, Соляев ЮО 2012 Моделирование зависимостей физико-механических характеристик от параметров микро- и наноструктуры полимерных композиционных материалов *Наука и образование: электронное научно-техническое издание* **6** 38 [Lurie S, Mironov Y, Nelyub V, Borodulin A, Chudnoy I, Buyanov I and Solyaev Y Simulation Dependency physical - mechanical characteristics from parameters and nanostructures polymer composite *Materials Science and Education: Electronic scientific - technical edition* (in Russian)]

[10] Malysheva G 2014 Predicting the endurance adhesive joints *Polymer Science-Series D* **7**(2) 145–147

[11] Malysheva G and Bodrykh NV 2011 Hot-met adhesives *Polymer Science-Series D* **4**(4) 301–303

IOP Conf. Series: Materials Science and Engineering **153** (2016) 012003 doi:10.1088/1757-899X/153/1/012003

Analysis of mechanical properties anisotropy of nanomodified carbon fibre-reinforced woven composites

A N Ruslantsev[1,2], Ya M Portnova[3], L P Tairova[2] and A M Dumansky[1,2]

[1]Institute of Machine Science named after A.A. Blagonravov of the Russian Academy of Sciences. 4 Maly Khariton'evsky per., 101990 Moscow, Russia

[2] Bauman Moscow State Technical University. 5 2nd Baumanskaya str, 105005 Moscow, Russia

[3] JSC "ORPE "Technologiya" named after A.G.Romashin". 15 Kievskoye Shosse, 249031 Obninsk, Kaluga Region, Russia

E-mail: andreiruslantsev@gmail.com

Abstract. The polymer binder cracking problem arises while designing and maintaining polymer composite-based aircraft load-bearing members. Some technological methods are used to solve this problem. In particular the injection of nanoagents can block the initiation and growth of microscopic cracks. Crack propagation can also be blocked if the strain energy release is not related with fracturing. One of the possible ways for such energy release is creep. Testing of the anisotropy of the woven carbon fibre reinforced plastic elastic characteristics and creep have been conducted. The samples with different layouts have been made of woven carbon fibre laminate BMI-3/3692 with nanomodified bismaleimide matrix. This matrix has a higher glass transition temperature and improved mechanical properties. The deformation regularities have been analyzed, layer elastic characteristics have been determined. The constitutive equations describing composite material creep have been obtained and its parameters have been defined. Experimental and calculated creep curves have been plotted. It was found that the effects of rheology arise as the direction of load does not match the direction of reinforcing fibres of the material.

1. Introduction

The unique physical and mechanical properties of polymer composites allow their wide use in the rocket and aerospace structures, but it is necessary to solve a number of issues related to the strength, durability, rigidity of these materials. Carbon fibre reinforced plastics (CFRP) are one of the most common materials used in aviation and space technology, they have high stiffness and strength of the material but at the same time they have a significant anisotropy of the properties, tendency to the damage accumulation, cracking, which forces engineering designers to apply an additional safety ratio. Some technological methods are used to solve cracking problem. In particular the injection of nanoagents can inhibit the initiation and growth of microscopic cracks. Crack propagation can be also blocked if the strain energy release is not related with fracturing. One possible way for such energy release is to take advantage of time-dependent effects.

It is important to analyse the patterns of the deformation processes and develop the computational and experimental methods predicting CFRP resistance to the deformation and destruction, depending on the structural and technological factors. In this paper the model predicting the creep of equally strong woven CFRP BMI-3/3692 with nano-modified bismaleimide matrix under shear loading in the layer plane is proposed.

The time-dependent effects appear under shear loading of CFRP, whereas during deformation along and across the direction of reinforcing, the rheological effects are negligible.

In [1, 2] experimental studies of CFRP deformation are conducted, and the impact of creep is studied. The change in the viscoelastic component is defined using the power law: $\varepsilon = \varepsilon_0(1+at^n)$, where $\varepsilon_0 = \sigma/E$. The essential disadvantage of this approach is the inability to take into account the history of the load [3]. The dependency between creep curves and strain rates is shown on Fig. 1 [3].

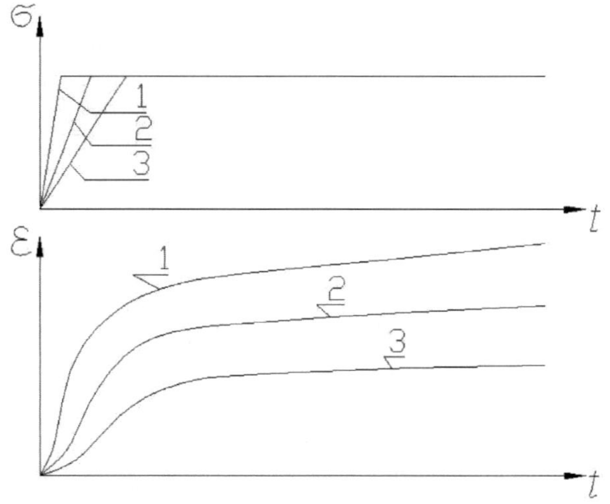

Figure 1. Creep curves obtained at different strain rates at an early stage

In [4-7] experimental studies based on carbon, glass and organic fibre reinforced plastics are conducted. It is shown that the most significant time-dependent effects arise during shear loading in the layer plane. In these works the same approach describing time-dependent effects is used. In [8] for the approximation of the experimental dependence Maxwell model is selected.

In [9-14] the relations of the linear hereditary mechanics proposed by Rabotnov are used. They have the form:

$$\varepsilon = \frac{\sigma}{E}(1 + K^*),$$

where K is the creep kernel:

$$K^*\sigma(t) = \int_0^t K(t-\tau)\sigma(\tau)\, d\tau.$$

Rabotnov fractional exponential function, the Abel kernel, Prony series [15] can be used as a creep kernel.

The hereditary approach is the most common and appropriate form of the relationship between stress and strain under variable loads [9]. This approach also makes possible the identification of a number of related functions characterizing the material. Thus, the expression for the description of the relaxation can be written as: $\sigma = E(1-R^*)\varepsilon$, where R^* is the relaxation kernel having the following interrelation $(1+K^*)(1-R^*) = 1$, $R^* = K^*/(1+K^*)$.

Paper [16] is devoted to the analysis of the time-dependent properties of CFRP and determination of the hereditary type constitutive equations of the of laminated CFRP KMU-4L. The rheological properties of the tensile specimens with layup [±40°], [±50°] are studied experimentally. These properties are determined primarily by the rheological properties under shear loading. Testing of the samples with the layup [±40°] under time-varying loads makes it possible to determine the time-dependent relations [17]. It has been found that this relationship can be represented as a hereditary type constitutive relation with Abel kernel [4].

2. Description of the experiment

The samples were cut from the 14-layer twill weave plate using computer marking and diamond wheel cutting with CNC machines. All layers were oriented in the same direction. Stress/strain curves for flat tensile specimens were obtained under loading in the Instron test machine. The strains along and across the longitudinal axis of the sample were measured using strain gages (Fig. 2) and strain gauge station SIIT-4. Test specimens were produced with longitudinal axis having an angle of 0°, 30°, 45° and 90° to the direction of the base (Fig. 3). The cross-sectional area of the samples was $2 \times 18 = 36$ mm^2.

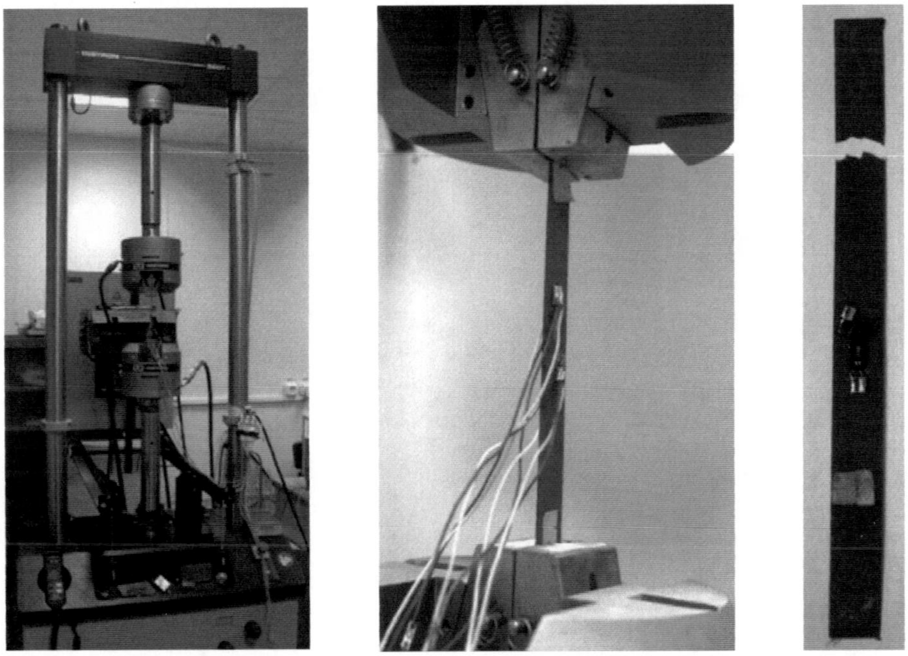

a) b) c)

Figure 2. a) The test machine; b) The sample in the grips; c) The sample after the test

Samples of each cutting direction were loaded at a speed corresponding to the quasi-static load with a subsequent discharge. Loading rate for each sample was varied in the 2-3 times. The final loading was carried out until the destruction. Three strain gauges were pasted onto each sample: two in the longitudinal direction and one in the transverse direction.

The sample 0° was loaded to stress 220 MPa, 440 MPa and 660 MPa, with subsequent discharge (Fig. 4), and then to the destruction that occurred at 836 MPa. The sample 90° was

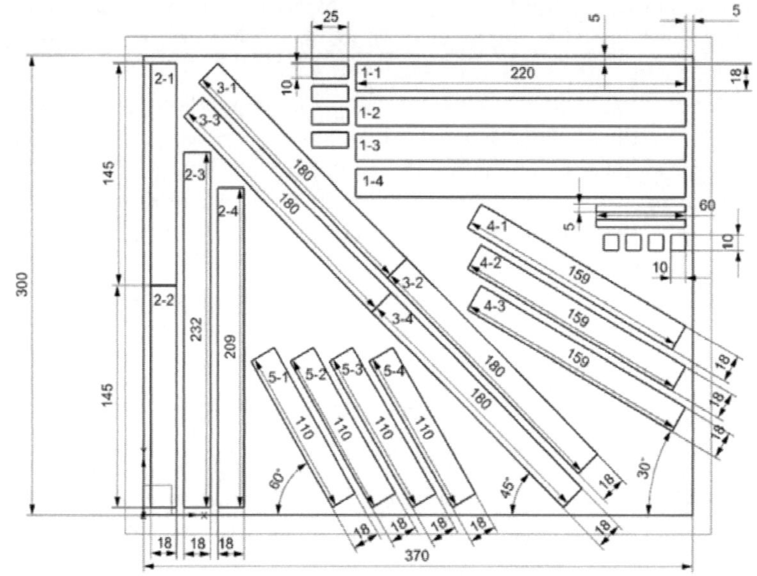

Figure 3. Nesting scheme

loaded to 75, 150, 300 and 450 MPa (Fig. 5), and then until destruction which occurred at 694 MPa. Strain diagrams were close to linear, elastic characteristic values are almost independent of the level of maximum stress and loading rate. These experiments enabled the identification of the following Young's moduli of the layer: $E_1 = 84$ GPa, $E_2 = 66$ GPa, $\nu_{12} = 0.04$.

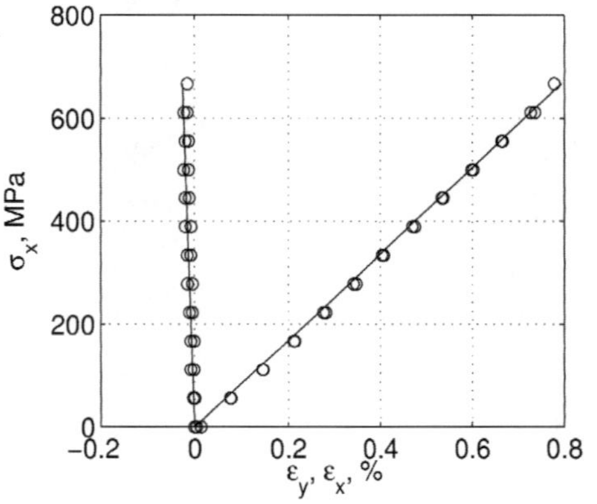

Figure 4. The experimental stress-strain curves of the sample 0°

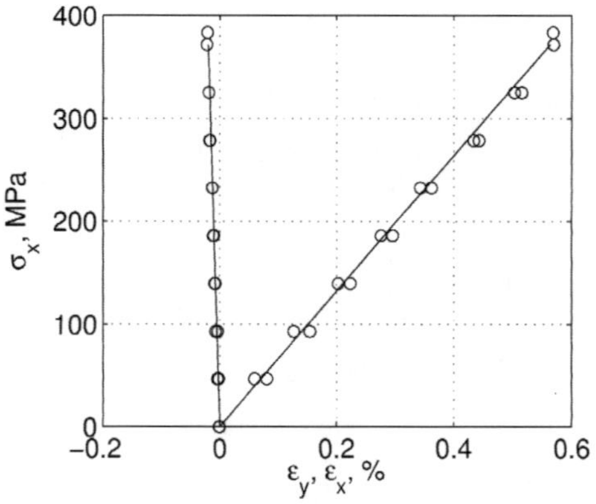

Figure 5. The experimental stress-strain curves of the sample 90°

The 30° sample was loaded to 100 MPa and 150 MPa at different speeds, then until the failure (222 MPa), Fig. 6. The 45° sample was loaded to 56, 111 and 167 MPa at different speeds, and then until the destruction (196 MPa). The stress/strain diagrams of these samples are close to linear for stress not higher than 0.4 of the destructive stress and become nonlinear when approaching the breaking load (Fig. 7). The form of the diagrams depends slightly on the strain rate.

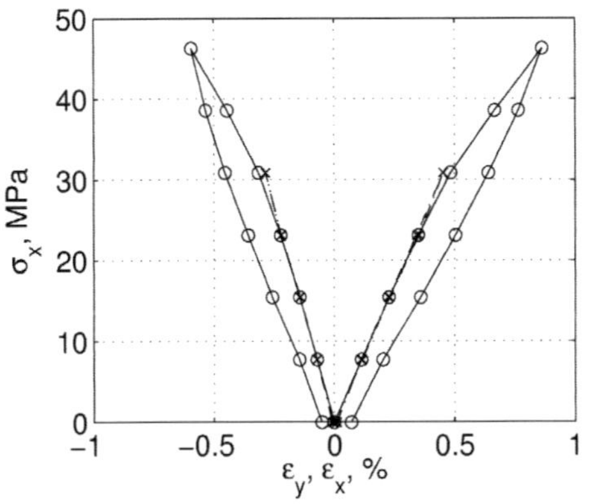

Figure 6. The experimental stress-strain curves of the sample 30°

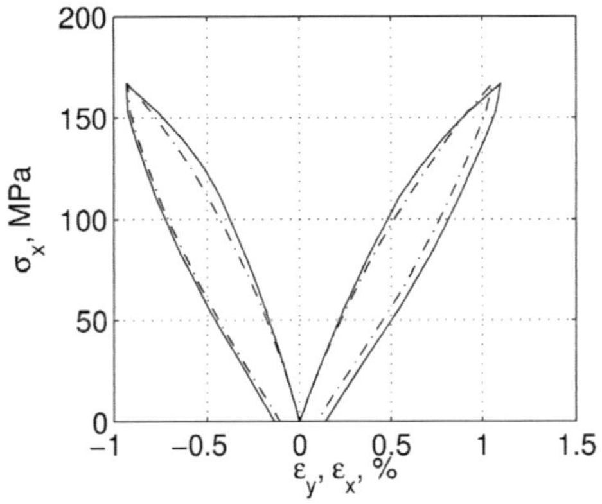

Figure 7. The experimental stress-strain curves of the sample 45°

The shear modulus was identified by solving the inverse problem for the linear part of the stress-strain curve for the specimens cut at angles 30° and 45°, $G_{12} = 7.5$ GPa.

The other samples cut at 45° were tested for creep. The longitudinal and transverse strains have been registered throughout the test. The first sample was loaded to 172 MPa in 15 seconds and held at this stress for 75 seconds, after that the failure occurred (Fig. 8). The maximum longitudinal strain before failure was 1.9%. Fig. 8 shows the dependence of the stress and strain on the time for this sample. Another sample was loaded up to 158 MPa in 15 seconds, and then for 1.7 hours was kept at this load without visible signs of destruction. The maximum longitudinal strain registered was 1.8%.

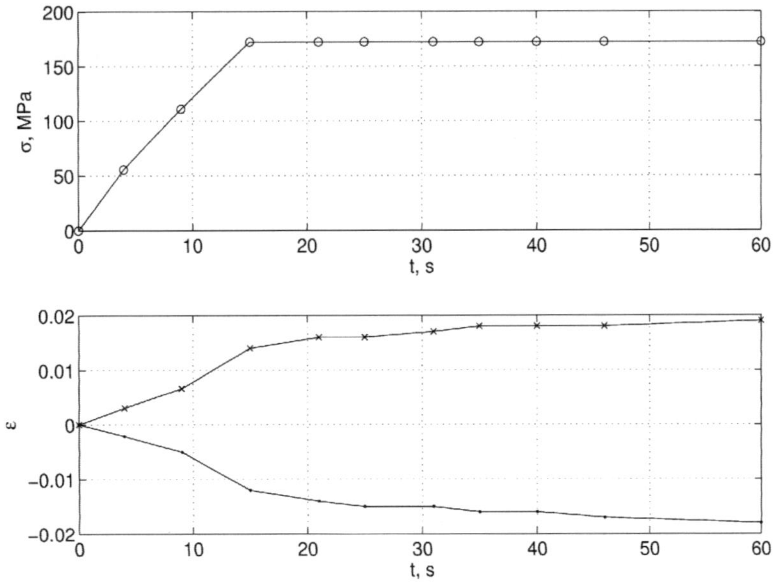

Figure 8. The experimental stress and strain curves of the sample cut at an angle of 45°

3. Model Description

It was shown [15] that the dependence between the stress and the strain can be represented as a hereditary type constitutive relation:

$$\gamma_{12} = \frac{1}{G_{12}^0}(1 + K^*)\tau_{12}.$$

Using the expression for the resolvent, an expression for the stresses calculated by the known history of strain was obtained [5]:

$$\tau_{12} = G_{12}^0(1 - R^*)\gamma_{12}.$$

Constitutive relations for the layer can be written in the following matrix form:

$$\begin{Bmatrix} \sigma_1 \\ \sigma_2 \\ \tau_{12} \end{Bmatrix} = \begin{bmatrix} \dfrac{E_1}{1 - \nu_{12}\nu_{21}} & \dfrac{\nu_{21}E_1}{1 - \nu_{12}\nu_{21}} & 0 \\ \dfrac{\nu_{21}E_1}{1 - \nu_{12}\nu_{21}} & \dfrac{E_2}{1 - \nu_{12}\nu_{21}} & 0 \\ 0 & 0 & G_{12}(1 - kR^*) \end{bmatrix} \begin{Bmatrix} \varepsilon_1 \\ \varepsilon_2 \\ \gamma_{12} \end{Bmatrix},$$

$$\{\sigma_{12}\} = [G_{12}^0 - G^0 R^*]\{\varepsilon_{12}\},$$

where $[G_{12}^0]$ is the stiffness matrix, $[G^0]$ is the matrix taking into account creep effect,

$$[G^0] = \begin{bmatrix} 0 & 0 & 0 \\ 0 & 0 & 0 \\ 0 & 0 & 1 \end{bmatrix}.$$

The stiffness matrix of the package can be calculated as:

$$[G_{xy}] = [T][G_{12}][T]^T = [G_{xy}^0] - [\tilde{G}]R^*,$$

where $[G_{xy}^0] = [T][G_{12}^0][T]^T$ is the stiffness matrix of the package without taking into account the time-dependent properties, $[\tilde{G}] = [T][G^0][T]^T$ is the correctional matrix, $[T]$ is the rotation matrix,

$$[T] = \begin{bmatrix} c^2 & s^2 & -2sc \\ s^2 & c^2 & 2sc \\ sc & -sc & c^2 - s^2 \end{bmatrix},$$

where $s = \sin(\theta)$, $c = \cos(\theta)$, $\theta = 45°$ in this case.

To obtain the compliance matrix it is necessary to invert the stiffness matrix:

$$[S_{xy}] = [G_{xy}]^{-1} = [G_{xy}^0 - \tilde{G}R^*]^{-1} =$$

$$= \left[[G_{xy}^0]\left([I] - [S_{xy}^0][\tilde{G}]R^*\right)\right]^{-1} = \left[[I] - [S_{xy}^0][\tilde{G}]R^*\right]^{-1}[S_{xy}^0],$$

where $[S_{xy}^0] = [G_{xy}^0]^{-1}$ is the compliance matrix of the package without taking into account time-dependent properties. Let $[A] = [S_{xy}^0][\tilde{G}]$. After diagonalisation we finally have $[A] = [Q]\ \mathrm{diag}(\lambda_i)\ [Q]^{-1}$, where $\mathrm{diag}(\lambda_i)$ is a diagonal matrix of A eigenvalues, and columns of $[Q]$ are the corresponding right eigenvectors.

Thus, after transformation, the following expression was finally obtained:

$$[S_{xy}^*] = [Q]\ \mathrm{diag}\left(1 - \lambda_i R^*(\mu - \lambda_i)\right)\ [Q]^{-1}[S_{xy}^0].$$

Constitutive equations for the package are written in the following form

$$\{\varepsilon(t)\} = [S^*_{xy}]\{\sigma_{xy}(t)\},$$

where $[S^*_{xy}]$ is a compliance matrix, $\sigma_{xy}(t)$ is a function describing the loading history.

A satisfactory approximation of the experimental data may be obtained using the sum of exponential functions as a kernel [15]. We have determined the parameters of the kernel by minimizing the discrepancy between the calculated and experimental values of the strain:

$$R(t) = 5.0 - 4.7\left[1 - \exp\left(-\frac{t}{10}\right)\right].$$

Calculated and experimental creep curves of CFRP with the layup of 45° are shown in Fig. 9.

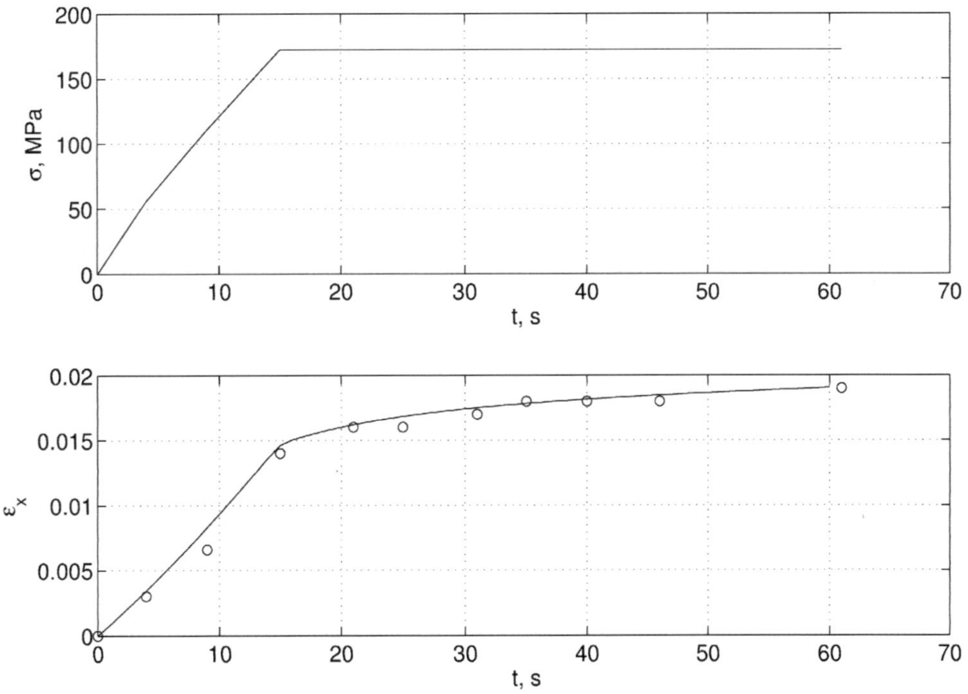

Figure 9. The calculated creep curve of CFRP and the experimental points of the 45° sample.

4. Conclusions

The analysis of mechanical properties anisotropy of woven CFRP with nano-modified high-temperature polymer matrix has been done.

Quasi-static tests have been conducted on specimens cut at different angles from the plate. The elastic and strength characteristics of the material in the longitudinal and transverse directions have been determined.

Creep tests have been conducted to identify the rheology effects. A hereditary model has been obtained, and the parameters of the model have been determined.

A satisfactory agreement between the calculated and experimental data has been shown.

References

[1] Schapery R A 1974 Viscoelastic behavior and analysis of composite materials *Mechanics of Composite Materials* **2** pp 86-168
[2] Deng S, Li X and Weitsman Y 2003 Time-Dependent Deformation of Stitched T300 Mat/Urethane 420 IMR Cross-Ply Composite Laminates *Mechanics of Time-Dependent Materials* **7** pp 41-69
[3] Koltunov M A 1976 Creep and relaxation (Moscow: Vyshaya shkola) p 277
[4] Kawai M, Masuko Y and Sagawa T 2006 Off-axis tensile creep rupture of unidirectional CFRP laminates at elevated temperature *Composites: Part A* **37** pp 257-269.
[5] Yao Z, Wua D, Chen C and Zhang M 2013 Creep behavior of polyurethane nanocomposites with carbon nanotubes *Composites: Part A* **50** pp 65-72
[6] Yahyaei-Moayyed M and Taheri F 2011 Experimental and computational investigations into creep response of AFRP reinforced timber beams *Composite Structures* **93** pp 616-628
[7] Sokolov E A 1980 The possibility of predicting the creep properties of the laminate organoplastic unidirectional fibre reinforced materials *Mechanics of Composite Materials* **1** pp 142-147
[8] Du Y, Yan N and Kortschot M 2013 An experimental study of creep behavior of lightweight natural fibre-reinforced polymer composite / honeycomb core sandwich panels *Composite Structures* **106** pp. 160-166
[9] Rabotnov Yu N, Papernik A Kh and Stepanych E I 1971 Nonlinear creep of fibreglass TS8/3-250 *Mechanics of polymers* **3** pp 391-397
[10] Rabotnov Yu, Papernik A Kh and Stepanych E I 1974 The relationship between the characteristics of fibreglass and instant creep curve *Mechanics of polymers* **4** pp 624-628
[11] Balevicius R and Marciukaitis G 2013 Linear and Non-linear creep models for a multi-layered concrete composite *Archives of civil and mechanical engineering* **13** pp 472-490
[12] Janson J O, Dmitrienko I P and Zelin V I 1983 Prediction of creep deformation organoplastic unidirectionally reinforced by the results of quasi-static tests *Mechanics of Composite Materials* **4** pp 610-613
[13] Maksimov R D and Plume E Z 1982 Prediction of creep unidirectional fibre reinforced plastic with termoreologicheski simple structural components *Mechanics of Composite Materials* **6** pp 1081-1089
[14] Plume E Z 1985 Comparative analysis of creep unidirectional composites reinforced with fibres of different types *Mechanics of Composite Materials* **3** pp 431-436
[15] Tamuzs V, Andersons I, Aniskevich K, Jansons Yu and Korsgaard J 1998 Creep and damage accumulation in orthotropic composites under cyclic loading *Mechanics of Composite Materials* **34** pp 447-460
[16] Dumansky A M and Tairova L P 2007 The prediction of viscoelastic properties of layered composites on example of cross ply carbon reinforced plastics *World Congress on Engineering* **2** pp 1346-1351
[17] Rabotnov Y N 1980 Elements of hereditary solids mechanics (Moscow: Mir) p 387

IOP Conf. Series: Materials Science and Engineering **153** (2016) 012004 doi:10.1088/1757-899X/153/1/012004

Finite element analysis of 4ENF testing of stitched carbon/epoxy laminate

Jonny Herwan[1], Atsushi Kondo[1, 2], Satoshi Morooka[1] and Naoyuki Watanabe[1]

[1] Department of Aerospace Engineering, Tokyo Metropolitan University, 6-6 Asahigaoka, Hino-shi, Tokyo 191-0065, Japan
[2] e-Xtream Engineering, Materials Modelling Company, 1-23-7 Nishishinjuku, Shinjuku-ku, Tokyo 160-0023, Japan

E-mail: jonny-herwan@ed.tmu.ac.jp

Abstract. Stitching technique improves both mode I and mode II delamination toughness significantly by reducing the opening and sliding displacement, respectively. However, measuring mode II delamination behaviour of stitched composites using three point end notched flexure test (ENF) is facing a serious problem where the premature failure of specimen occurs prior to delamination propagation. This work is addressed to conduct numerical study of four points end notched flexure (4ENF) test of stitched composites. Four models were simulated including unstitched unidirectional, unstitched multidirectional laminate, stitched multidirectional laminate with and without tabs on top and bottom of specimen. Furthermore, stitch threads were modelled using spring connectors with user define force-displacement relationship. Load-displacement and crack propagation length vs. displacement curves were plotted. Stress distributions across the thickness were also presented to recognize the possibility of specimen damage during the test. The results showed that simple 4ENF test seems to be quite enough for evaluating mode II critical energy release rate (G_{IIC}) at the initial crack length of 35mm. However to evaluate R-curves (G_{II} vs. crack propagation length) of stitched multidirectional laminate, tabbed 4ENF test specimen is recommended.

1. Introduction

Despite of excellent strength and stiffness to weight ratio of carbon fibre reinforced plastics (CFRP), the delamination between plies become a serious issue of conventional laminates. Stitching is one of potential 3-Dimensional reinforcement by inserting threads in the thickness direction during preform fabrications.

The improvement of mode I delamination toughness of stitched laminates have been evaluated successfully using double cantilever beam (DCB) test [1, 2]. However, evaluation of mode II delamination properties using three points end notched flexure (ENF) is facing a difficult problem. The laminate is failed prior to delamination propagation. To overcome this problem, tabbed end notched flexure (TENF) test has been conducted [3, 4]. Embedded plates (tabs) at top and bottom side of the specimen require a modification of data reduction method [3]. In case of stable crack propagation, compliance calibration method could be directly applied for TENF test.

Four points end notched flexure (4ENF) test is a simple test with stable crack propagation[5]. Many data points could be recorded with stable crack propagation, so that R-curve (energy release rate vs. crack propagation) could be obtain using one specimen. This work is addressed to conduct feasibility study of using 4ENF test for mode II delamination properties of stitched composites. Both normal 4ENF and tabbed 4ENF specimen of stitched multidirectional laminate are investigated numerically using commercial finite element software (ABAQUS). The maximum normal stresses are used to predict whether the specimens failed before the delamination reach mid of the span. Furthermore, R-curve of stitched composite was also predicted based on compliance calibration method.

2. Modelling techniques

2.1. Modelling of 4ENF test

The schematic set-up for Tabbed 4ENF testing is shown in figure 1 and related finite element model is described in figure 2. The fixture dimensions refer to the work reported by Davies et.al [6]. It is important to be noted that center loading pin has capability to be rotated so that the load is kept balanced during the test. Similar set-up for normal 4ENF test is modelled by omitting the tabs. In this study, the fixture is also considered in the finite element analysis because the relatively high load used during the test for stitched laminate. Therefore the fixture is also modelled as 3-dimensional geometry instead of a simple rigid shell. The laminate consist of 20 layers of carbon fibers with lay-up order of [+45/90/-45/0/0/+45/90/90/-45/0]s. The mechanical property of laminae is calculated based on homogenization method using in house codes [7] and summarize in table 1. The thickness of each layer is 0.205 mm, and total thickness of specimen is 6.35 mm including aluminum tabs and adhesive bonding. Four types of finite element models are developed, called unstitched unidirectional, unstitched multidirectional laminate, stitched multidirectional laminate with and without tabs. Both of unidirectional and multidirectional laminate consist of 20 layers as described above. The delamination propagates at 0°/0° interface for all models.

The crack tip is positioned exactly after the first stitch row with the distance of 35 mm from the lower-right support pin as shown in figure 1. This crack tip position (in between first and second stitch rows) is important in the experiment to capture effect of stitching on fibers volume fraction [4].

Table 1. Mechanical property of laminae

E_1 (GPa)	E_2 (GPa)	E_3 (GPa)	G_1 (GPa)	G_2 (GPa)	G_3 (GPa)	υ_{12}	υ_{13}	υ_{23}
158.6	4.7	4.7	3.8	3.8	2.7	0.33	0.33	0.47

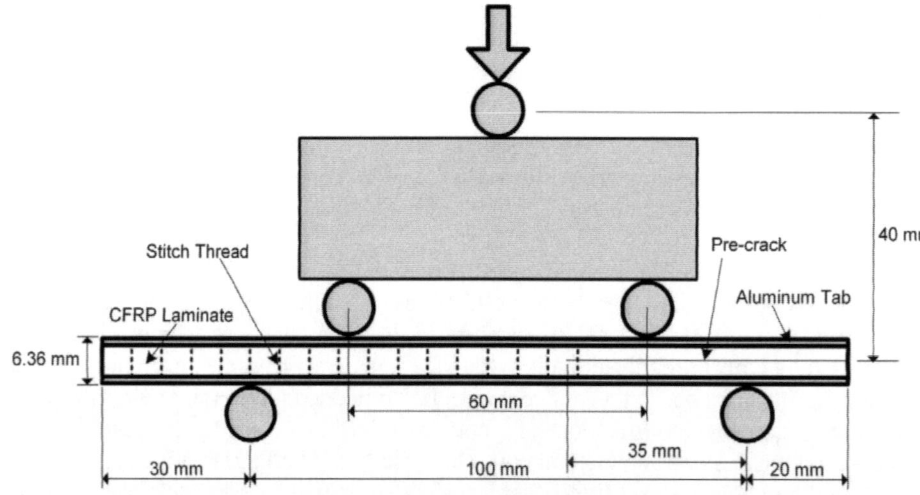

Figure 1. Set-up of tabbed 4ENF test.

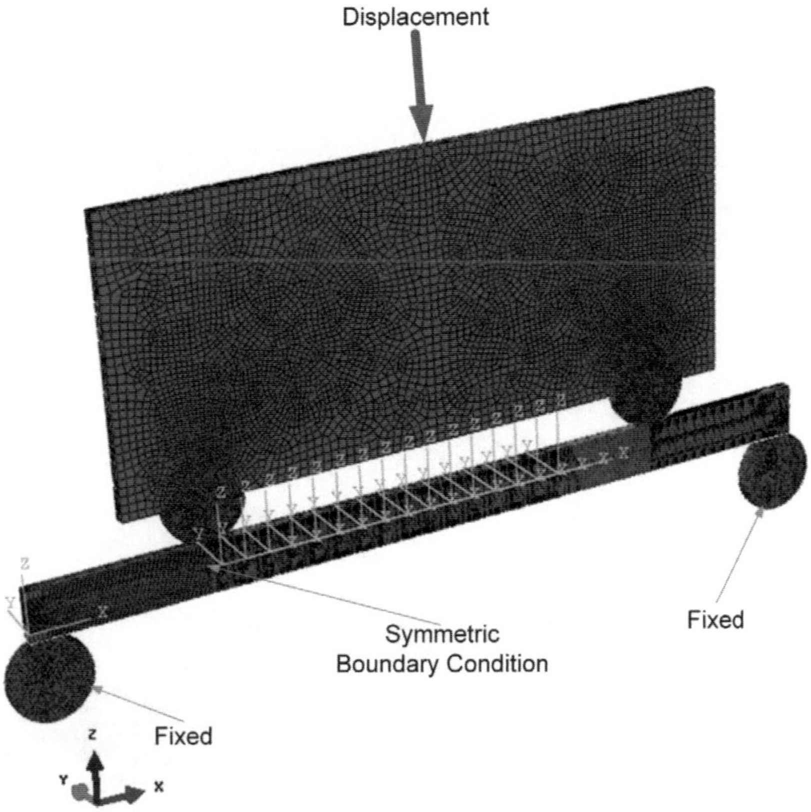

Figure 2. Finite element model of tabbed 4ENF test of stitched laminate.

2.2. Modeling of stitch threads

The stitch threads are modeled by spring connector elements. Mechanical properties of spring connector are defined using inter-laminar shear tested (IST) of single stitched laminate (figure 3) [8]. All nodes in stitch thread areas at interlaminar surface of upper and lower laminate are constraint by multi point constraints (MPCs) as shown in figure 4. The diagonals of MPCs area are measured from actual specimens. Furthermore spring connector element connects the center node of MPC area. Figure 5 presents a typical result of single stitched laminate of previous work [8] that adopted in modeling 4ENF test.

Figure 3. Single stitched laminate under pure shear testing.

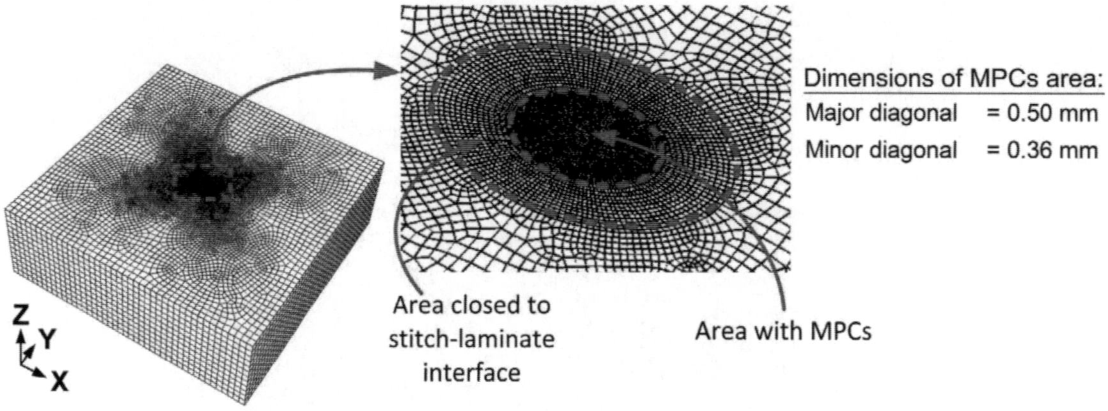

Figure 4. Modelling of single stitched laminate[8].

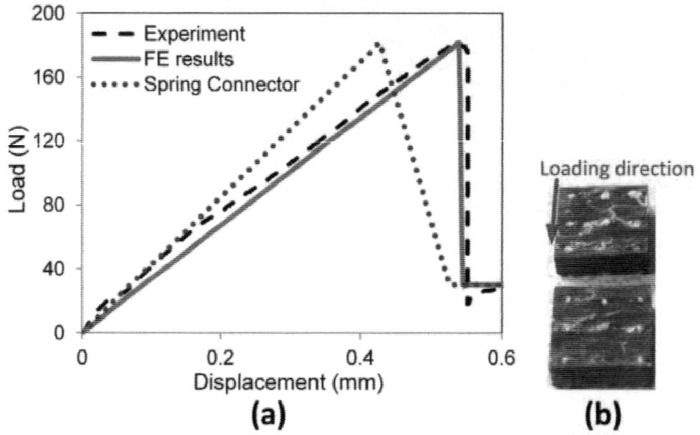

Figure 5. Single stitched laminate under pure shear testing.

2.3. Delamination criterion

In order to simulate delamination propagation, cohesive zone elements are used and inserted at mid-layer (figure 6). Initial delamination happened whenever a quadratic stress failure criterion (equation (1)) is satisfied.

$$\left\{\frac{(\sigma_I)}{\sigma_I^o}\right\}^2 + \left\{\frac{\sigma_{II}}{\sigma_{II}^o}\right\}^2 + \left\{\frac{\sigma_{III}}{\sigma_{III}^o}\right\}^2 = 1 \qquad (1)$$

where $\sigma_{I,\ II,\ III}$ are stress component under pure mode I, II, and III, meanwhile $\sigma^0_{I,\ II,III}$ are critical interfacial strength for each kind of loading mode.

Furthermore, damage (delamination) growth is controlled by mixed mode fracture energy criterion (equation (2)) with exponential softening fracture-based method. Mode I and mode II energy release rates are obtained from experiment while mode III is assumed to be the same with mode II. The cohesive zone parameters are listed in table 2 which are obtained from previous work [8].

$$\left\{\frac{G_I}{G_I^c}\right\}^\alpha + \left\{\frac{G_{II}}{G_{II}^c}\right\}^\alpha + \left\{\frac{G_{III}}{G_{III}^n}\right\}^\alpha = 1 \qquad (2)$$

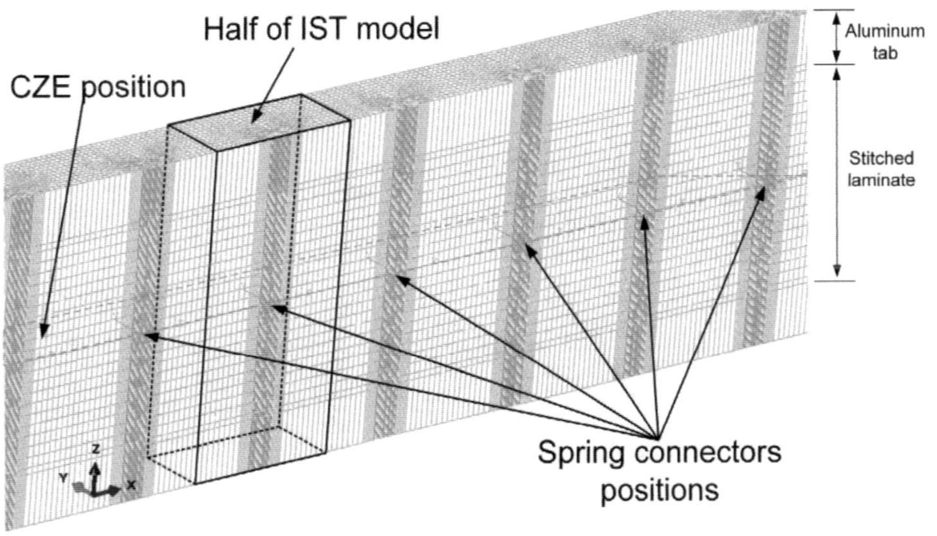

Figure 6. Tabbed 4ENF meshing technique.

Table 2. Cohesive zone parameters

K_0 (N/m^3)	σ_I^o (MPa)	σ_{II}^o (MPa)	σ_{III}^o (MPa)	G_I^c (J/m^2)	G_{II}^c (J/m^2)	α
4.25 x 10^{14}	40	70	70	450	800	1

2.4. Energy release rates prediction
Energy release rate is calculated from the simulation results based on compliance calibration method following equation (3):

$$G_{IIC} = \frac{mP^2}{2B} \tag{3}$$

where m is the slope of compliance vs. crack length, P is a certain load where delamination start to propagate, and B is the width of the specimen. To obtain the value of m, the compliance of the model are simulated at the crack length of 35, 38, 41, and 44mm. Noted that these crack length involved 1, 2, 3, or 4 active stitch threads at the delaminated area, respectively.

3. Results and discussion

3.1. Load displacement curve
Load-displacement curve of unstitched laminate models for unidirectional and multidirectional laminate are plotted in figure 7 and 8. The load-displacement curve of unidirectional laminate model showed higher stiffness compared to multidirectional one and need shorter deflection for crack tip to reach the centre of loading span. This could be happened because the delamination occurred at the same interface (0°/0°) and the mode II energy release rate (area under the curve) should be the same.

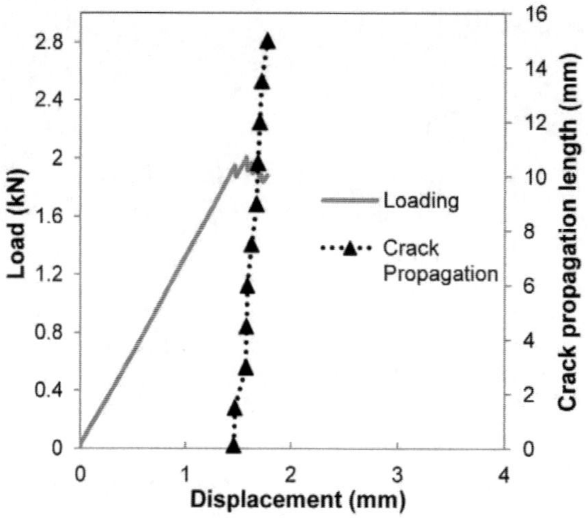

Figure 7. Load-displacement curve of unstitched unidirectional laminate.

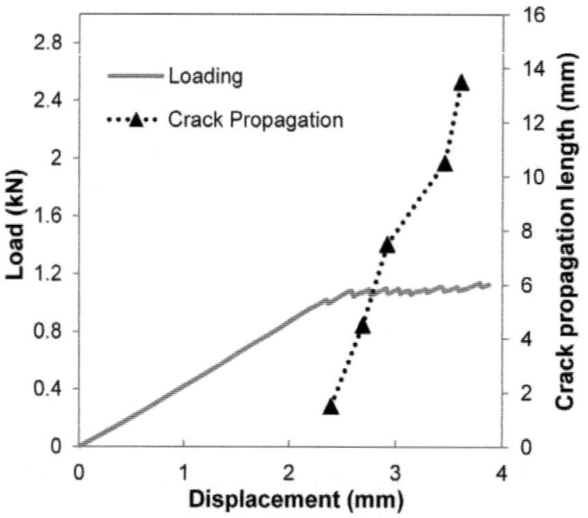

Figure 8. Load-displacement curve of unstitched multi-directional laminate.

Load-displacement curve of stitched laminate using 4ENF and T4ENF are plotted in figure 9 and 10. Both stitched laminate models are multidirectional laminate. On the other hand, stitched unidirectional laminate is not considered in this study because stitching process usually requires multidirectional in-plane fibre orientation, thus the preform can be stretched in many directions to minimize fibre waviness. Aluminium tabs at T4ENF specimen increased the stiffness but decreased the deflection/displacement needed to propagate the crack till mid of span. Both curves also showed that the load still increased during crack propagation which indicate that the more active stitch threads at delaminated area, the more load could be absorbed by the stitch threads.

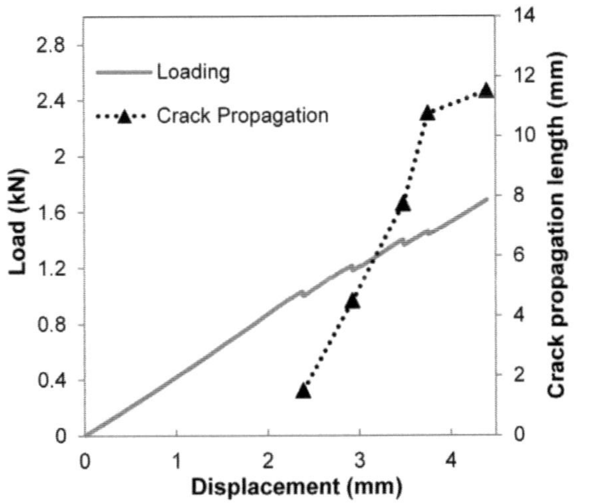

Figure 9. Load-displacement curve of stitched multidirectional laminate under 4ENF test.

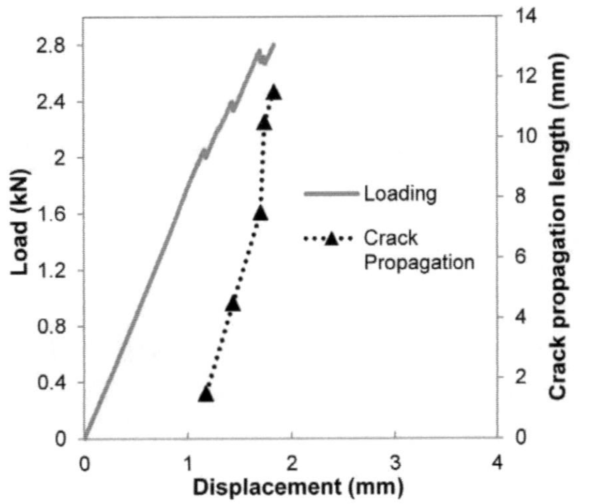

Figure 10. Load-displacement curve of stitched multidirectional laminate under Tabbed 4ENF test.

3.2. Stress distribution along the thickness direction

Figure 11-12 exhibit normal stresses (S_{11}) distribution of 4ENF and T4ENF simulation results at the end of simulation in Figure 9-10. It is important to be noted that actually the highest stress area is around stitch region at delaminated area, but it is excluded from the figures because the energy dissipated by damages around stitch threads are taken into account in G_{IIC} calculation. However, damage occurred in another region (except at interlaminar area) is prohibited during mode II delamination test.

In order to investigate the possibility of damage at each layer, the maximum normal stresses at the middle of loading span region are plotted in figure 13. It is clearly shown that aluminum tabs significantly reduced maximum normal stresses at each layer to become about one third of those in laminate without tabs. Additionally, since the yield strength of aluminium alloy 7075 T6 is about 450-500 MPa, it could also be ensured that there is no plastic deformation at aluminium tab until the end of simulation. Therefore, T4ENF specimen is recommended for mode II delamination testing of stitched composites.

The use of standard 4ENF specimen seems still acceptable for limited purpose such as measuring G_{IIC} at initial crack length of 35 mm. To understand the maximum condition where standard 4ENF can be used, further study is required to investigate matrix crack during the test simulation which strongly depend on matrix properties.

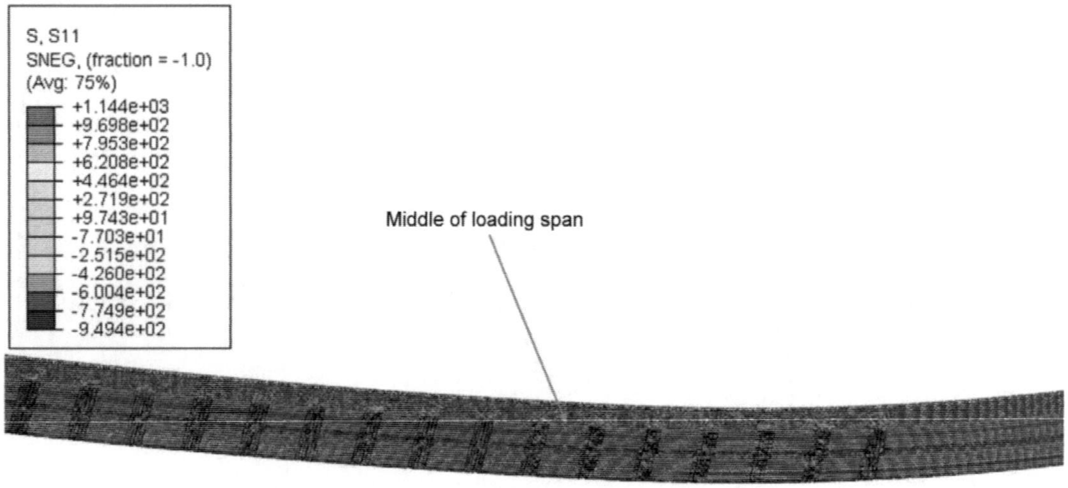

Figure 11. Stress distribution (S_{11}) in 4ENF simulation.

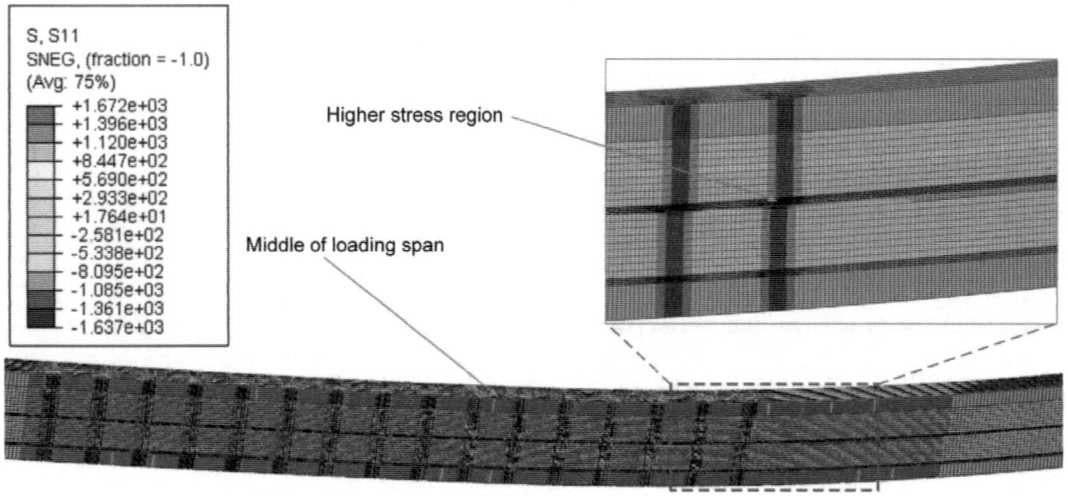

Figure 12. Stress distribution (S_{11}) in tabbed 4ENF simulation.

3.3. Predicting of energy release rate

A typical compliance vs. crack length curve is plotted in figure 14. The slope (m) can be calculated from the curve as shown in the same figure. Furthermore energy release rates were calculated using equation 2 and plotted in figure 15. Figure 15 exhibited energy release rate of stitched laminate is twice the unstitched one ($0.88 kJ/m^2$). However, G_{II} values are still increasing continuously and saturated value could not be achieved. The more crack propagate, the more active stitch thread at delaminated region. Saturated region could be achieved if the stitch thread starts to be broken. From

IOP Conf. Series: Materials Science and Engineering **153** (2016) 012004 doi:10.1088/1757-899X/153/1/012004

displacement data of each spring connector, the maximum displacement is about 0.2 mm which is much longer to be broken. Increasing deflection of specimen, until stitch thread broken, has to be done carefully considering matrix crack and fiber rupture. The thickness of the aluminum tabs could be increased as well, but one has to check shear stress at the interface of aluminum and laminate. Thicker tab create higher shear stress at the tabs-laminate interfaces that prone to be de-bonding.

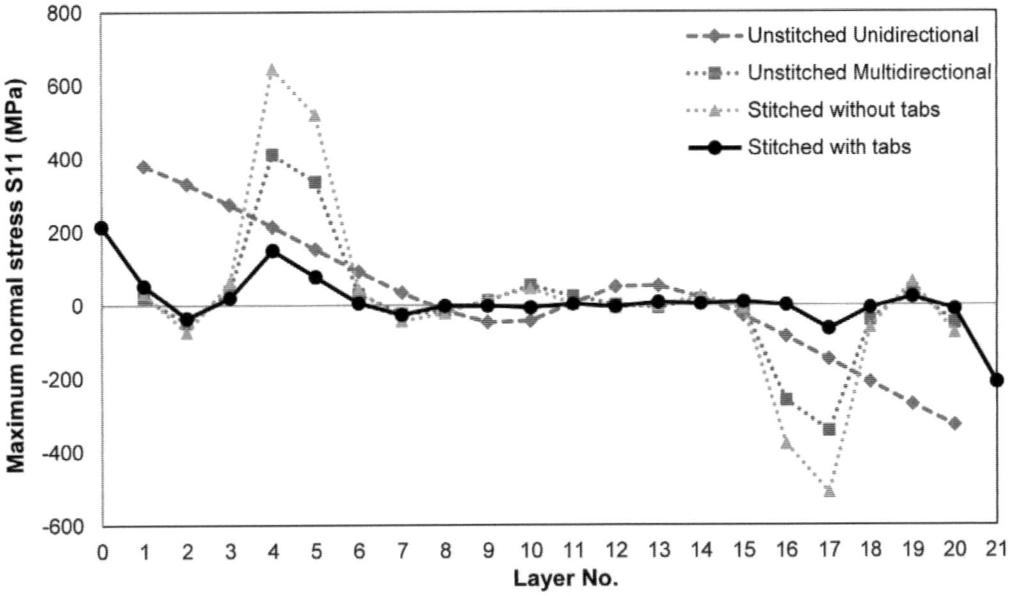

Layer No	0	1	2	3	4	5	6	7	8	9	10	11	12	13	14	15	16	17	18	19	20	21
Orientation	LowerTab	45	90	-45	0	0	45	90	90	-45	0	0	-45	90	90	45	0	0	-45	90	45	UpperTab

Figure 13. Maximum normal stress distribution across the thickness.

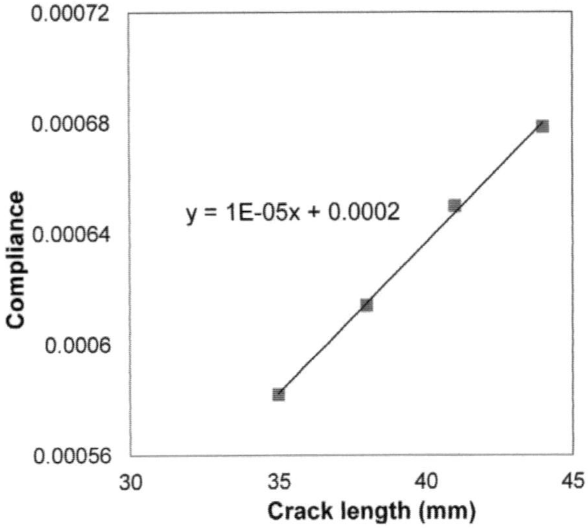

Figure 14. Compliance vs crack length.

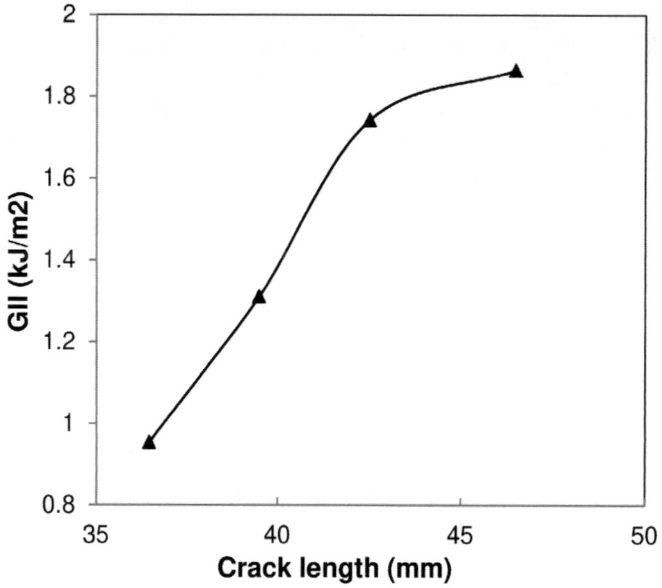

Figure 15. Energy release rate vs. crack length (R-curve).

4. Conclusion

A finite element analysis of 4ENF and T4ENF test has been conducted. In order to obtain energy release rates vs crack length (R-curve), T4ENF has to be used to avoid matrix crack. Predicted energy release rates of stitched laminate are twice the value of unstitched one. Higher energy release of stitched laminate could be achieved with longer crack length, but one has to ensure that there is no matrix crack occurred. It is also possible to increase thickness of aluminium tab to reduce more normal stress (S11) in each layer of laminate. In case of increasing the thickness of aluminium tab, the shear stress between tabs and laminate should be carefully controlled to be not more than the critical interfacial stress.

Acknowledgments

Authors wishing to acknowledge Tokyo Metropolitan Government for the financial support through the program of Human Asian Resource Fund and Asian Network of Major Cities 21 (ANMC-21) Project.

References

[1] Dransfield K A, Jain L K and Mai Y-W 1998 On the effects of stitching in CFRPs—I. Mode I delamination toughness *Composites Science and Technology* 58 815–27

[2] Tan K T, Watanabe N, Sano M, Iwahori Y and Hoshi H 2010 Interlaminar Fracture Toughness of Vectran-stitched Composites - Experimental and Computational Analysis *Journal of Composite Materials* 44 3203–29

[3] Wood M D K, Sun X, Tong L, Luo Q, Katzos A and Rispler A 2007 A New ENF Test Specimen for the Mode II Delamination Toughness Testing of Stitched Woven CFRP Laminates *Journal of Composite Materials* 41 1743–72

[4] Herwan J, Kondo A, Morooka S and Watanabe N 2014 Effects of stitch density and stitch thread thickness on mode II delamination properties of Vectran stitched composites *Plastics, Rubber and Composites* 43 300-8

IOP Conf. Series: Materials Science and Engineering **153** (2016) 012004 doi:10.1088/1757-899X/153/1/012004

[5] Martin R H and Davidson B D 1999 Mode II fracture toughness evaluation using four point bend, end notched flexural test *Plastics, Rubber and Composites* 28 401-6
[6] Davies P, Casari P and Carlsson L A 2005 Influence of fibre volume fraction on mode II interlaminar fracture toughness of glass/epoxy using the 4ENF specimen *Composites Science and Technology* 65 295-300
[7] Nasution M R E, Watanabe N, Kondo A and Yudhanto A 2014 Thermomechanical properties and stress analysis of 3-D textile composites by asymptotic expansion homogenization method *Composites Part B: Engineering* 60 387-91
[8] Herwan J, Kondo A, Morooka S and Watanabe N 2015 Finite element analysis of mode II delamination suppression in stitched composites using cohesive zone model *Plastics, Rubber and Composites* 44 390-6

IOP Conf. Series: Materials Science and Engineering **153** (2016) 012005 doi:10.1088/1757-899X/153/1/012005

Synthesis of a discrete-action thermo-bimetallic actuator with a tongue

A Nikolaeva[1], **AJ McMillan**[2], **S Gavriushin**[1]

[1] Bauman Moscow State Technical University, 5/1, 2nd Baumanskaya, Moscow, 105005, Russia.
[2] Wrexham Glyndwr University, Mold Road, Wrexham, LL11 2AW, UK.

nikolaeva@bmstu.ru

Abstract. The selection of suitable parameters, by experimental or intuitive processes for snap-through actuation of a bimetallic actuator at a prescribed temperature is an extremely time-consuming task. This paper describes a new methodology for the optimization of a discrete action thermo-bimetallic actuator with a tongue. This methodology makes it possible to solve the optimization task with higher efficiency. The requirement is to find optimal parameters values so that the actuator will make a snap-through at a given temperature. The constrained optimization task was performed using an evolutional algorithm and surrogate modelling and this was coded in Matlab. Functional relationships between the criteria and parameters were not set explicitly, but they were calculated using finite element method, each simulation of which was performed in Abaqus.

1. Introduction

Discrete action actuators are used in industry. In particular they can be used as thermo-sensitive electrical switches for overload protection of electric circuits, for important elements temperature control and regulation, in optical components, as a manipulator for object transportation in micro-systems.

Dome plate bimetallic actuators are well known [1]. They change curvature direction, making a snap-through at a critical temperature [2-4]. Despite the ease of fabrication, they have several disadvantages. For example, their temperature of a snap-through is unstable and the component experiences high stresses, in some cases exceeding the elasticity limit. In the latter case, the value of the critical temperature changes over as a result of the evolution of the zone of plastic deformations. As a result the service life of such actuators, relays and switches is limited. Another disadvantage is the emergence of cracks on the actuator surface after multiple switches, leading to actuator failures.

Another well-known type of actuators is that of actuators with a tongue [5]. Their effective displacement is higher compared with the dome actuators, so they can ensure required contact forces and avoid bounce. Required contact force ensuring makes it possible to transform the mechanical displacement of the actuator into the proper performance of the following part of the micro-electro-mechanical system (MEMS). Due to the displacement paucity, contacts can "bounce" upon closure before coming to a full rest and providing unbroken contact, which is inadmissible in electric circuits.

Bimetallic actuator parameter optimisation using current state-of the-art optimisation and finite element modelling (FEM) analysis software (such as ANSYS, Abaqus, iSight) is an extremely time-consuming task, as it requires multiple meshing after geometrical design parameter variations. The formal use of these programs for the analysis of bimetallic actuators deformation processes without correct design strategy can often be unsuccessful (especially in the case of deformation with a snap-through).

The snap-through behaviour and large deflections of thin-walled shells has been studied by many authors [6-12]. Other recent research papers [13-14] describe computational algorithms for non-linear deformation. Optimization plays a key role in the design of actuators [15-18], however, most MEMS design optimization (exploration) methods depend either on analytical / behavioural models or on time consuming numerical simulations. Surrogate modelling techniques have been introduced to integrate generality and efficiency [19].

In the present paper, a technique is presented for determining geometry parameters for a discrete action micro-actuator, operating under snap-through. The actuator consists of two regions: hemispherical shell with a central hole and tongue configuration (Figure 1). The actuator is comprised of two layers. The parameters are shown in figure 1. It is required to find optimal parameters values so that the actuator will make a snap-through at a given temperature. This constrained optimization task was performed using an evolutional algorithm and surrogate modelling [19] and implemented in MATLAB R2013a [20]. Functional relationships between the criteria and the parameters are not set explicitly, but they are calculated using a finite element method [21], each simulation of which is performed in Abaqus finite element software [22].

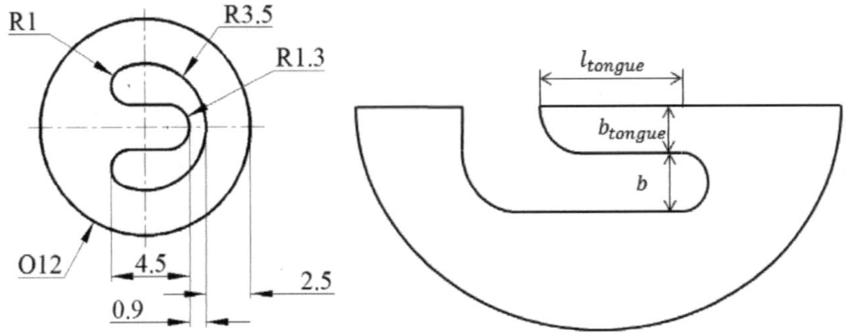

Figure 1. Actuator - geometry and parameters for the optimization task.

2. The method adopted

The hemi-spherical thin-walled micro-actuator is designed to perform mechanical switching. Let us suppose that its deflection is required to take a given value d^* when the critical temperature T^* is applied. The elastic characteristic of the micro-actuator which includes a snap-through mechanism and the deformed shape of the actuator are shown schematically in figure 2.

The shell curvature changes after applying temperature. This part of the process is illustrated by the DA part of the curve. The shell instantly changes the direction of the deflection at the point A (the upper critical point) – the snap-through at point A. Then the shell continues its deformation from the point B of the stable part of the elastic characteristic. The deflection of the shell decreases if the pressure decreases (BC) until the point C (lower critical point) and the CD snap-through.

The initial data for the research is given in table 1. The required outcome of the optimisation process is the selection of values $l_{\text{tongue}}, b_{\text{tongue}}, b$, subject to the value of critical temperature T^* being equal to the customer's requirements. The acceptable ranges of variation of these parameters are also given. The functional relationships between the criteria and parameters are not set explicitly, but are calculated using Finite Elements in Abaqus as described below.

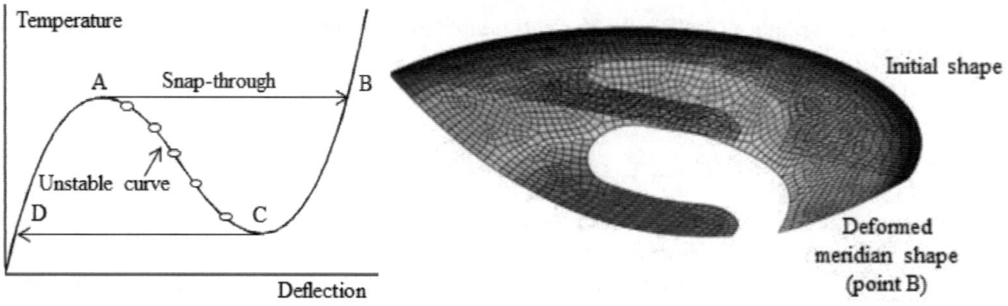

Figure 2. Schematic of the discrete elastic response and deformed shape of the actuator.

The shell is modelled using shell elements. The material of the shell is linearly elastic. The shell is simply-supported around the edge and is loaded by temperature. The analysis is performed using the parameter continuation algorithm arc-length method for a problem with large displacements. In this method the length of the arc [23] along the elastic characteristic curve is used as a parameter in the parameter continuation method, rather than displacement in the displacement - controlled mode or temperature in the temperature-- controlled mode. This makes it possible to overcome computational stability difficulties. The evolutional algorithm and surrogate modelling method implemented in Matlab are used for the solution of the optimization problem [21].

There is one objective function, specified as

$$F_1 = |T - T^*| \tag{1}$$

Table 1. Problem specification data

Parameter		**Value**	
Length of the tongue	l_{tongue}	$1.5 - 4 \times 10^{-3}$ m	
Width of the tongue	b_{tongue}	$1.3 - 2 \times 10^{-3}$ m	
Hole parameter	b	$1.2 - 2 \times 10^{-3}$ m	
Desired temperature	T^*	40°C	
		Layer 1	**Layer 2**
Shell thickness	h	1×10^{-4} m	1×10^{-4} m
Young's modulus	E	190×10^9 Pa	150×10^9 Pa
Poisson's ratio	ν	0.3	0.3
Thermal expansion coefficient	α	18×10^{-6}, 1/°C	1×10^{-6}, 1/°C

At each iteration, new values for the variables were substituted into the Abaqus file, and the finite element analysis was repeated. The calculated values for the critical temperature, T, were extracted from the Abaqus results database. The Optimizer continues to test different parameter values until the value of T matches that of T^* to within the tolerance requirements. Penalty functions were used to implement the constrained optimization.

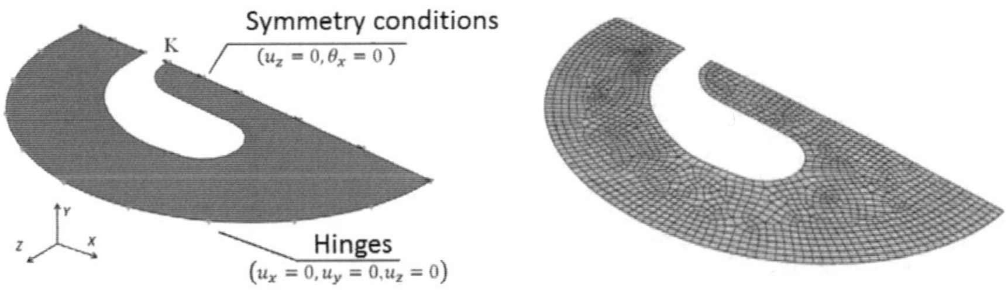

Figure 3. Mathematical and finite element models of the actuator

3. Evolutional algorithm and surrogate modelling

The evolutionary computation algorithm imitates the biological mechanisms of evolution to approximate the global extremum problem [15]. The main feature of the evolutionary algorithm (EA) is the use of individual populations that are processed by a set of operators (*crossover, mutation, selection*) and are evaluated using the fitness function. The goal of the operators is to find candidates with higher fitness function values [16]. The fitness function shows "how well" a candidate succeeded and determines the probability of its survival. The process of solving the problem consists of generating new populations and testing them: the candidate with the higher fitness function has a higher chance of being saved and being used as the "parent" when creating the next candidate solutions. First, populations (each population consists of individual specimens – actuators, which are characterized by a vector of parameters) are randomly generated. Then, the procedure of *crossover* is carried out: different candidates shares their binary strings parts to create new child populations (binary genetic algorithm [24]). The purpose of this step is the exchange of information by parents. If the useful information is combined, the functional adaptation of the new candidate populations is more likely to have high value. The next step is the *mutation*. The purpose of mutation is to present new information to the population to provide a global search. New populations are generated after the mutation stage. Candidates with a higher fitness function receive a greater likelihood of promoting successful generations on the *selection* stage. The best solution is displayed, if the stopping criterion is satisfied, *i.e.* the maximum number of populations is reached, otherwise, a new iteration starts.

Surrogate models are used to replace models which require time-consuming calculations? For example the computationally demanding FEM snap-through analysis. The purpose of the surrogate model is to significantly increase the efficiency of the optimization by replacing time-consuming computational steps with simpler models which are sufficiently representative the true values. In this study, the critical temperature for new parameter values is determined using a surrogate model of the surface of equilibrium states without extra running the finite-element program.

4. Results

The optimal parameters of the actuator are shown in the table 2. The elastic characteristic for the optimal geometry is shown in figure 4. The average time cost is 3.6 hours (wall clock time) which is five times quicker than optimization task solution using a standard differential evolution algorithm.

5. Discussion

Surrogate model-based methods are used in actuator design optimization to combine generality and efficiency. An initial random sampling is first carried out. The black-box surrogated model is prepared to approximate the performance of the actuator using the sampled design variables as the input and the critical temperature via numerical simulations as the output. It is used to replace computationally expensive numerical simulation model in the optimization process. The developed algorithm made it possible to achieve the results comparable with methods which directly embed numerical simulations to a standard evolutionary algorithm for actuator optimization.

Table 2. Optimal parameters

Parameter	Value
l_{tongue}	3.87×10^{-3} m
b_{tongue}	1.39×10^{-3} m
b	1.28×10^{-3} m

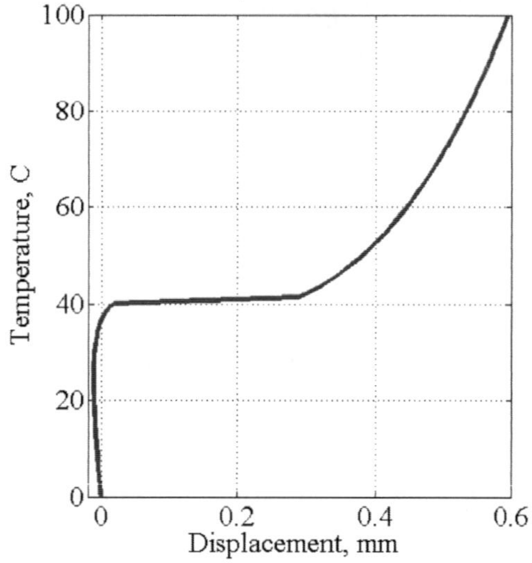

Figure 4. Elastic characteristic of an actuator with optimized geometry

The surrogate modelling method is Gaussian Process, providing an approximation model for the MEMS performance using existing training data points. For off-line surrogate model-based methods, prediction results are used as results based on real simulation in the optimisation process. Thus, its accuracy needs to be verified, or wrong convergence may happen; however, the surrogate model here is an on-line model. Promising candidates using prescreening methods are selected and are then verified by finite element analysis. In other words, the surrogate model only serves for looking for "promising" solutions, but the prediction result itself is not used as the performance. The surrogate model assisted evolutional algorithm used here is a state-of-the-art method [25].

The efficiency of the optimization process by using surrogate modelling was improved by up to five times. Actuator optimization problems with several design variables and without any initial solutions were solved.

6. Conclusion

A technique for determining geometric parameters for a discrete action actuator with a tongue, to translate a prescribed critical temperature into deflection has been described. The three geometry variables considered were the tongue length, tongue width and the hole parameter. The mechanical analysis was performed using Finite Element Analysis to obtain the temperature of a snap-through. The optimization process was performed using an evolutional algorithm and surrogate modelling in Matlab program. The proposed methodology followed has been demonstrated to be effective, a number of opportunities for methods development have been identified, and these will form the basis for further more challenging design optimization studies. Compared with other state-of-the-art methods, the proposed technique showed clear advantages in efficiency and accuracy.

References

[1] Moorhead J O 1962 Patent (UK) 1031827 *Thermally responsive electrical switch.*

[2] Gavriushin S S, McMillan A J and Nikolaeva A S 2015 Discrete action micro-actuator optimization *IOP Conference Series: Materials Science and Engineering* **75.**

[3] Gavriushin S S, McMillan A J, Nikolaeva A S and Podkopaeva T B 2015 Расчет перспективных конструкций актюаторов [Calculation of actuator promising designs] *Известия высших учебных заведений. Машиностроение [Proc Higher Educational Institutions. Machine building]* **8** (665) 73-78.

[4] Gavriushin S S, McMillan A J and Podkopaeva A S 2014 Синтез микроактюатора дискретного действия по заданным функциональным параметрам [Discrete action micro-actuator synthesis for required parameters] *Известия высших учебных заведений. Машиностроение [Proc Higher Educational Institutions. Machine building]* **1** (646) 55-60.

[5] Taylor J C 1978 Patent (USA) 4160226 *Snap-acting thermally responsive actuators*.

[6] Timoshenko S P and Gere G M 2009 *Theory of elastic stability* 2nd ed Dover, New York 544. Timoshenko S P, 1971, *Устойчивость стержней, пластин и оболочек. [Stability of rods, plates and shells]* Nauka, Moscow [in Russian].

[7] Voitovich I I 1989 *Математические проблемы нелинейной теории пологих оболочек [Mathematical problems of the nonlinear theory of shallow shells]* Nauka, Moscow [in Russian].

[8] Feodos'ev V I 1969 Осесимметричная эластика сферической оболочки [Axisymmetric elastic spherical shell] *Prikladnaia matematika i mekhanika* [*J Applied Mathematics and Mechanics*] **33**(2) 280-286 [in Russian].

[9] Grigoliuk E I 1953 Тонкие биметаллические оболочки и пластины [Bimetallic thin shells and plates] *Inzhenernyi sbornik* [*Engineering collection*] **17** 69-120 [in Russian].

[10] Novozhilov V V 2010 *Теория тонких оболочек [The theory of thin shells]*, Saint Petersburg University, Saint Petersburg [in Russian].

[11] Reissner E 1950 On axisymmetrical deformations of thin shells of revolution *Proc. Sympos. Appl. Math.* **3** 27-52.

[12] Popov E P 1948 Явление большого перескока в упругих системах и расчет пружинных контактных устройств [Snap-through in elastic systems and contact devices springs calculation] *Inzhenernyi sbornik* [*Engineering collection*] **5** 62-92 [in Russian].

[13] Bich D H and Tung H V 2011 Non-linear axisymmetric response of functionally graded shallow spherical shells under uniform external pressure including temperature effects *Int. J. Nonlinear Mechanics* **46**(9) 1195-1204.

[14] Li Q S, Liu J and Tang J 2003 Buckling of shallow spherical shells including the effects of transverse shear deformation *Int. J. Mechanical Sciences* **45**(9) 1519-1529.

[15] Liu B, Fernández F V, and Gielen G 2014 *Automated design of analog and high frequency circuits,* Springer, Berlin, Heidelberg.

[16] Coello C A, Lamont G B and van Veldhuizen D A 2007 *Evolutionary algorithms for solving multi-objective problems*, Springer Science+Business Media, New York.

[17] Cortez P, 2014, *Modern Optimization with R*, Springer, Switzerland.

[18] Venkataraman P 2002 *Applied Optimization with MATLAB Programming.* Wiley-Interscience, New York.

[19] Ong Y S, Nair P B and Keane A J 2003 Evolutionary optimization of computationally expensive problems via surrogate modelling *Amer. Inst. Aero, Astron. J.* **41**(4) 687-696.

[20] Gilat A 2014 *MATLAB An introduction with Applications* 2nd Ed John Wiley & Sons.

[21] Zienkiewicz O C 1977 *The Finite Element Method* McGraw-Hill, New York.

[22] ABAQUS 2011 ABAQUS Documentation, Dassault Systèmes, Providence, RI, USA.

[23] Crisfield M A 1981 A fast incremental/iterative solution procedure that handles snap through *Computer* & Structures **13**(1) 55-62.

[24] Fogel D B 2006 *Evolutionary computation: toward a new philosophy of machine intelligence*, John Wiley & Sons, New York.

[25] Liu B, Zhang Q and Gielen G 2014 A Gaussian Process Surrogate Model Assisted Evolutionary Algorithm for Medium Scale Expensive Black Box Optimization Problems *IEEE Transactions on Evolutionary Computation* **18**(2) 180-192.

IOP Conf. Series: Materials Science and Engineering **153** (2016) 012006 doi:10.1088/1757-899X/153/1/012006

Optimization of the rheological properties of epoxy resins for glass and carbon reinforced plastics

Pyi Phyo Maung, G Malysheva, and I Romanova

Bauman MSTU, Moscow, Russia

pyaephyo@mail.ru

Abstract. Vacuum assisted resin transfer moulding (VARTM) offers advantages such as simplicity, low cost of consumables, and the ability to carry out the impregnation process and curing without using expensive equipment and tooling. In the VARTM process, rheological properties of resin have a critical impact on the impregnation and curing process. In this article, the experimental results of viscosity are presented, including the glass transition temperature, and the tensile and bending strength of the epoxy binders with the amine hardener, which depend on the quantity of its active solvent composition. The active solvent used is diethylene glycol. It shows that for an increase in the content of the active solvent, a reduction in the viscosity and a reduction of the glass transition temperature and strength occurs. The optimum composition of the binder is selected by using the Pareto optimization criteria and the Cayley – Smorodinskaya method. By using the epoxy binder, the active solvent should not exceed 10-15% by weight. This approach helps to optimize the amount of active solvent added to the epoxy resins for the criterion of viscosity, strength, and heat resistance.

1. Introduction

Fibreglass and carbon reinforced plastic composite materials are widely used in various industries, including aerospace, aeronautics, automotive, marine and others [1]. Different types of pre-preg are traditionally used to form a composite structure; however, there are many disadvantages such as the expense of the technology and limited viability of pre-preg for small-scale production.

In recent years, different types of composite moulding process are widely used in the composite industries. Among them, vacuum Assisted Resin Transfer Moulding (VARTM), also referred to as infusion, has been developed as a variant of the traditional Resin Transfer Moulding (RTM) process to reduce the cost and design difficulties and to eliminate the pre-preg requirements for composite structure. In VARTM, the upper half of a conventional mould is replaced by a vacuum bag. This eliminates the need for making a precisely matched metal mould as in the conventional RTM process. The resin dampens the fibre, and the part is then cured inside the mould.

The main advantages of infusion technology are simplicity, low cost of consumables, the ability to carry out the impregnation process and curing without the use of expensive equipment and tooling.

The speed of the impregnation process is determined by the properties of the binder and woven fibre; however, in practice problems arise often when the designer is given the specific type of woven fabric. In this situation, all the properties of the given carbon woven fabric were taken account while choosing the binder.

The most widely used binders in the manufacturing of carbon fibre reinforced plastic structure are epoxy binders. The advantages of epoxy resin are good adhesion, high mechanical properties, low shrinkage and very good processing properties. The main technological disadvantage of epoxy resins is very high viscosity.

Various additives are used in the composition of epoxy resin to reduce the viscosity. Among them, active solvents are the most widely used as an additive; however, this leads to a reduction of heat resistance and a reduction of mechanical strength while adding the active solvent in composition to improve the dampening characteristic and rheological properties of other binders [2].

The aim of this work is to optimize the active solvent content in the epoxy resins for the criterion of viscosity, strength, and heat resistance.

2. Theoretical approach

A variety of indicators can be used as optimality criterion "h". In general, the optimality condition has the form

$$h_l(\mathbf{x}') < h_l(\mathbf{x}'') = \mathbf{x}' \succ_X \mathbf{x}'',$$
$$h_2(\mathbf{x}') > h_2(\mathbf{x}'') = \mathbf{x}' \succ_X \mathbf{x}'', \qquad (1)$$

where h_l and h_2 are two criteria that characterize the properties of the object of study; X is the solution set obtained; x' and x'' are solutions for first and second criteria; and $\succ$ indicates a preference for solution x'compared with solution x''in the set X.

As a result, from the equation (1), the impregnation process is better and faster for smaller values of criterion h_l, and on the other hand, a higher value of criterion h_2 leads to better properties.

It is proposed to use h_l as a criterion of the binder viscosity index, because the lower the viscosity, the quicker and more accurately the pass process impregnates the woven fabrics.

As a criterion for h_2 (h_3, h_4, etc.) a variety of indicators may be used, such as the glass transition temperature (h_2), bending strength (h_3) and the tensile stress (h_4).

When using the multiple criteria, a correlation must be established between them using the correlation coefficient r. The correlation coefficient r is defined as follows

$$r_{1,2} = \frac{I_{1,2} - I_1 I_2}{\sqrt{[(I_{1,1} - I_1^2)(I_{2,2} - I_2^2)]}}$$
$$I_1 \cong \frac{1}{N}\sum_{i=1}^{N} h_1(A_i) \qquad (2)$$
$$I_{1,2} \cong \frac{1}{N}\sum_{i=1}^{N} h_1(A_i) h_2(A_i)$$

where $r_{1,2}$ is the correlation coefficient for criteria numbered 1 and 2; and A_i is the point in the area of parameters to be optimized.

3. Experimental approach

In this study, the properties of 8 epoxy binder compositions, each with different levels of active solvent content are investigated experimentally. Results are shown in table 1. The active solvent used is diethylene glycol (DEG).

After adding different amounts of active solvent, the viscosities of the binders were determined by Brookfield viscometer CAP-2000. The glass transition temperature was evaluated by differential scanning calorimetry, DSC 204 F1. These binders were filled in standard forms (GOST 12015) and after curing were tested for tensile (GOST 11262) and bending strength (GOST 4648). The experimental results for viscosity, glass transition temperature, tensile and bending strength are shown in table 2.

Table 1. Different diethylene glycol compositions of epoxy binder

Binder composition	Active solvent DEG content (% by weight)
1	0
2	1
3	5
4	10
5	15
6	20
7	25
8	30

Table 2. Properties of the binders

Binder composition	Viscosity (Pa s)	Glass transition temperature (°C)	Max. Tensile strength (MPa)	Max. Bending strength (MPa)
1	17	61	109	139
2	12.5	61	105	143
3	3.6	58	90	145
4	1.15	58	90	148
5	0.86	58	85	115
6	0.54	56	83	107
7	0.27	56	85	110
8	0.18	56	85	107

Table.3 The values of correlation coefficients

Criteria	The values of the correlation coefficient
Active solvent content - Viscosity	0.7861
Active solvent content - glass transition temperature	0.9091
Active solvent content - destructive bending stress	0.8734
Active solvent content - destructive tensile stress	0.8164
Glass transition temperature - Bending and tensile strength	0.9439

4. Results and discussion

Using equation (2) and the results in table 2, the correlation coefficients were determined, as shown in table 3.

Analysis of the values of correlation coefficients in table 3 show that the highest one is the criteria of the glass transition temperature – bending and tensile strength. This means that the increasing binder active solvent content leads to a decrease in the glass transition temperature and reduction in bending and tensile strength.

Since the viscosity reduction is a positive factor, this value can be directly included as one of the criteria for solving the optimization problem in terms of minimization.

The glass transition temperature is the second criterion for evaluating the effectiveness of the increase of active solvent content. Lowering this value has a negative impact because it reduces the operating temperature range.

Using the relative characteristics, optimality criterion changes are

$$\bar{h}_2 = \Delta h_2 = h_{2\max} - h_2 \tag{3}$$

The correlation coefficient of this pair is 0.9154, which could be assigned to the class of multi-criteria optimization problem [3].

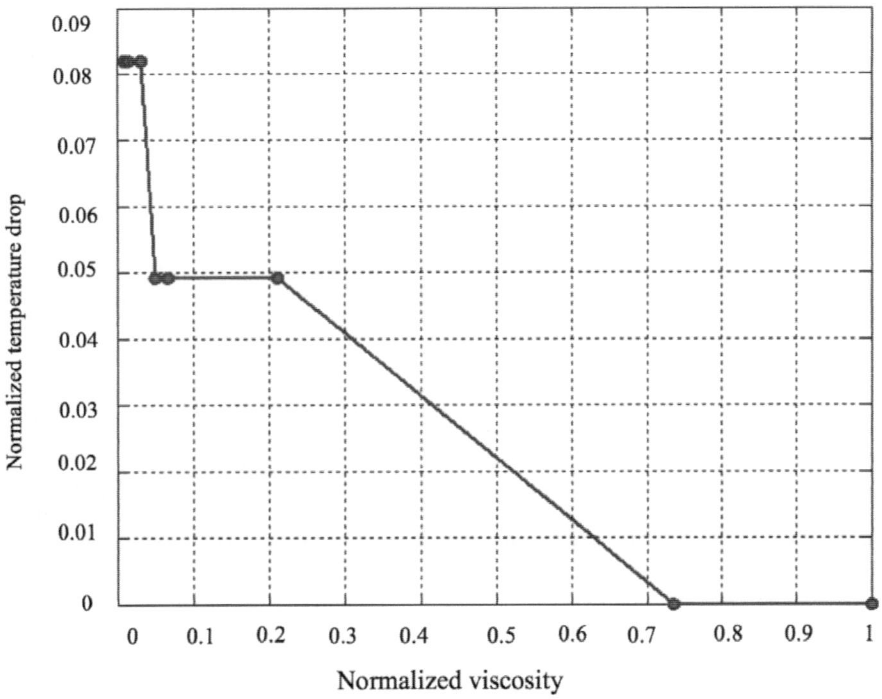

Figure 1. Dependence of normalized criteria on the viscosity and glass transition temperature drop

According to the Pareto criterion, optimization approach of the resin composition is

$$\begin{array}{l} \mathbf{x}', \mathbf{x}'' \in X, h_i(\mathbf{x}') \le h_i(\mathbf{x}''), i = 1,...m \\ \exists k \in \{1,2,...,m\}: \quad h_k(\mathbf{x}') < h_k(\mathbf{x}'') = \mathbf{x}' \succ_X \mathbf{x}'' \end{array} \tag{4}$$

The test criteria were further normalized,

$$\begin{array}{l} h_{1n} = h_1 / h_{1\max} \\ \bar{h}_{2n} = (h_{2\max} - h_2) / h_{2\max} \end{array} \tag{5}$$

The dependence of the normalized criteria is presented in figure 1. The points illustrate different binder variants.

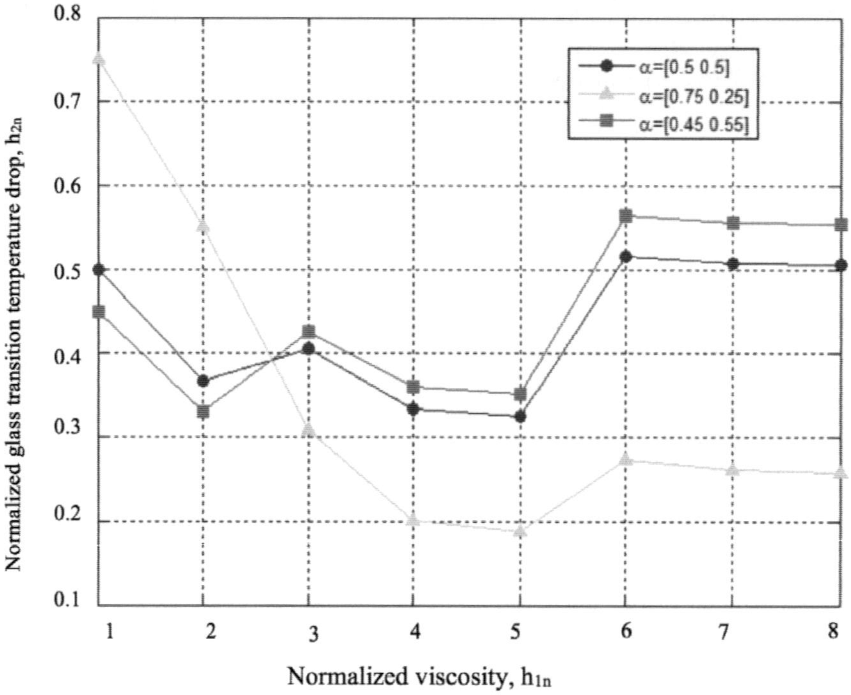

Figure 2. Dependence of the normalized criteria on the viscosity and glass transition temperature on the characteristics of active solvent

To obtain a unique solution, randomized strategies were used based on the information about the relative importance of criteria [4]. When choosing the optimal solution, the theorem of the vector criterion $h = (h_1, \tilde{h}_2)$ is used with weight coefficients $\{\alpha_i\}$. The corresponding generalized criterion calculated by the formula is thus

$$J(\mathbf{x}, \alpha) = \langle h(\mathbf{x}), \alpha \rangle = \sum_{i=1}^{m} \alpha_i h_i \tag{6}$$

where $\alpha = (\alpha_1, ..., \alpha_m)$ – vector of non-negative weight coefficient satisfies the equation,

$$\sum_{i=1}^{m} \alpha_i = 1 \tag{7}$$

As shown in figure 1, the points illustrate the different binder variants in table 1, which correspond to the values of the normalized viscosity. The normalized criteria of viscosity and glass transition temperature for different values of the weight coefficients is shown in figure 2.

Three weight coefficient values are presented in figure 2 with different colours, and the optimum compositions are compositions 4 and 5, i.e. DEG compositions of 10 or 15%.

Verification of the result was conducted using the theory of cooperative games [5]. For this optimization approach, the initial data was set at the maximum allowable values for each variable. The Cayley - Smorodinskaya solution is shown in figure 3.

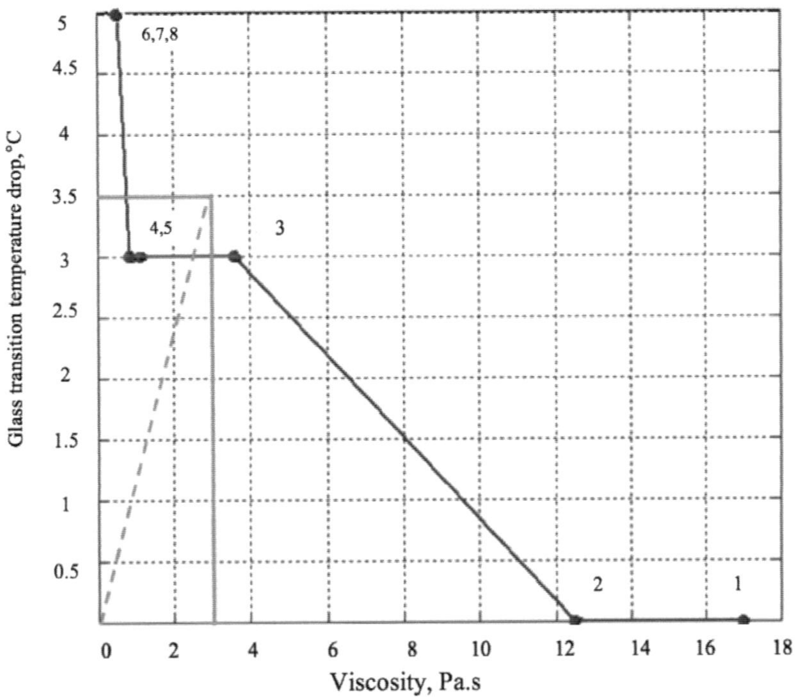

Figure 3. The Cayley-Smorodinskaya optimal solution for viscosity – temperature reduction

As follows from the data obtained, the optimum compositions are compositions 4 and 5, which corresponds to the previous solution.

5. Conclusions

The study shows the influence of the content of epoxy binder active solvent on viscosity change, glass transition temperature and mechanical properties after curing. It has been determined that by increasing the content of active solvent in the binder there is decrease of viscosity, which has a positive effect on the entire process of manufacturing the product; however, this method of viscosity reduction results in deterioration of heat resistance and mechanical strength. By using the Pareto optimization criteria, the best epoxy binder compositions were predicted to be those with DEG composition of 10 or 15%.

References

[1] Baurova N 2013 Microstructural Investigations of Surfaces of Destruction of Carbon Plastic *Polymer Science - Series D.* **6**(2) 246–249

[2] Malysheva G, Akhmetova E and Marycheva A 2014 Estimation of glass transition temp erature of poly sulfone -modified epoxy binders *Glass Physics and Chemistry* **40**(5) 543–548

[3] Lotov A and Pospelov I 2008 *Multicriteria decision making problems* (Moscow: Moscow State University Press)

[4] Chernorutsky I 2005 *Methods of decision-making* (S.Publishing House of Petersburg)

[5] Шварц ДТ 2013 Интерактивные методы решения задачи многокритериальной оптимизации. Обзор. Наука и образование. Электронный журнал **4** [Schwartz DT 2013 Interactive methods of solving the problem of multi-criteria optimization. *Overview. Science and education. Electronic journal*, in Russian]

A mathematical model of the inline CMOS matrix sensor for investigation of particles in hydraulic liquids

DV Kornilin and IA Kudryavtsev

Laboratory 16, Radioelectronic methods and devices for diagnostic of aircraft systems, Samara University, Samara, 443086, Russia.

E-mail: kornilin@mail.ru

Abstract. One of the most effective ways to diagnose the state of hydraulic system is an investigation of the particles in their liquids. The sizes of such particles range from 2 to 200 μm and their concentration and shape reveal important information about the current state of equipment and the necessity of maintenance. In-line automatic particle counters (APC), which are built into hydraulic system, are widely used for determination of particle size and concentration. These counters are based on a single photodiode and a light emitting diode (LED); however, samples of liquid are needed for analysis using microscope or industrial video camera in order to get information about particle shapes. The act of obtaining the sample leads to contamination by other particles from the air or from the sample tube, meaning that the results are usually corrupted. Using the CMOS or CCD matrix sensor without any lens for in-line APC is the solution proposed by authors. In this case the matrix sensors are put into the liquid channel of the hydraulic system and illuminated by LED. This system could be stable in arduous conditions like high pressure and the vibration of the hydraulic system; however, the image or signal from that matrix sensor needs to be processed differently in comparison with the signal from microscope or industrial video camera because of relatively short distance between LED and sensor. This paper introduces mathematical model of a sensor with CMOS and LED, which can be built into hydraulic system. It is also provided a computational algorithm and results, which can be useful for calculation of particle sizes and shapes using the signal from the CMOS matrix sensor.

1. Introduction

Most automatic particle counters that are built into hydraulic systems (in-line) are based on a single photodiode as a photo-sensor. This approach has advantages: the relatively small size of the sensor, low cost and reliability; however, it is not possible to distinguish particle shapes, which can be considered as important information about processes in entire hydraulic systems [1, 2]. In addition, photodiodes confine the sensitivity and maximum concentration limit. The present authors propose the use of a matrix sensor in order to deal with these problems. A matrix sensor consists of several relatively small photodiodes with integration scheme, digitizer and interface on the crystal. CMOS and CCD matrix sensors are widely used for obtaining pictures in different applications.

There are several mathematical models of the data provided by photoelectric particle sensors, which are based on the effect of illumination of the photodiode by LED [3–5]. All of these models utilize the light blocking method [6, 7]. In this method the estimation of particle sizes is calculated by relating the change of light intensity to the illuminated cross-sectional area of the particle in the sensor volume. This change in light intensity is estimated using a photodiode and some dedicated electronics. Figure 1 reflects the typical composition of such sensors. They consist of glass tube coated by nontransparent material with a small aperture in its centre. Light from the LED goes through the aperture to the photo-sensor.

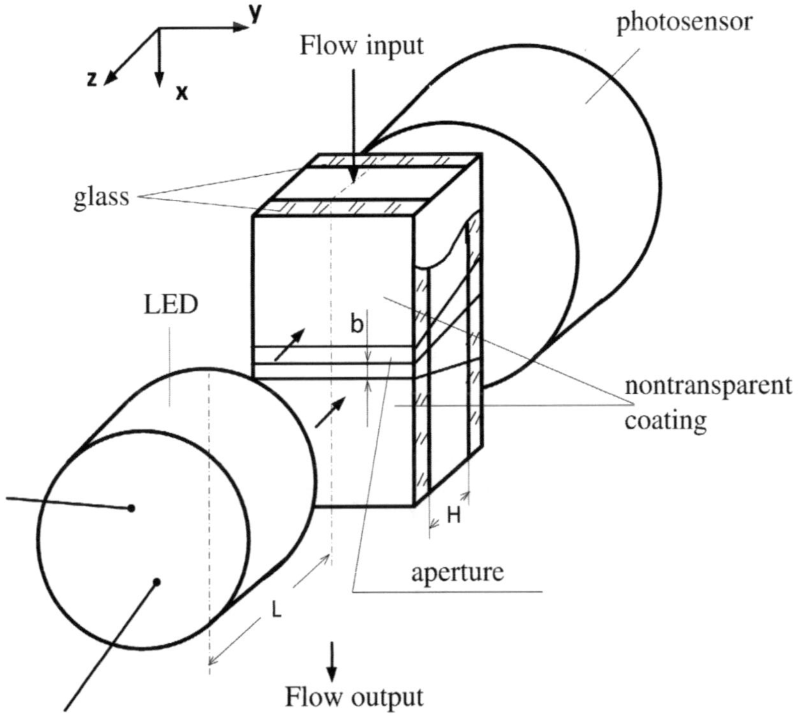

Figure 1. Typical composition of the optoelectronic part of the inline sensor.

In [5], the model based on geometrical optics has been described. This model can be used for the determination of sensitivity, and estimation of the accuracy influenced by particles position inside the volume. The sensor proposed by the authors is based on the matrix CMOS photo-sensor, instead of using a single photodiode, and it has quite a similar composition but with the exception of the necessity of an aperture. In this case, the mathematical model should be different in comparison with model in [5], because of the signal that is obtained from CMOS matrix and the absence of the aperture.

2. The mathematical model of the inline CMOS matrix sensor

The mathematical model described in this paper is based on the assumption that the LED and CMOS matrix sensor can be represented as equivalent planes that consist of several unit elements as the relatively small pixels, as illustrated in figure 2. It is certainly appropriate for the CMOS and can be applicable for the LED. The LED consists of a lens and a relatively small crystal in its focus. This composition can be considered optically as a significantly larger area at the higher distance, emitting through the lens aperture [5].

The LED is introduced as the plane with dimensions of A_S and B_S. The Cartesian coordinates of lower left corner are denoted as (x_{S0}, y_{S0}, H). The LED unit element has dimensions dx and dy with coordinates depending on its position that can be described as (x_S, y_S, H). The size of matrix sensor is represented by A_R and B_R. The lower left matrix corner has Cartesian coordinates (0, 0, 0). Each pixel has dimensions Δx and Δy with coordinates (x_m, y_m, 0). A particle that is placed between the LED and the matrix has coordinates (x_p, y_p, z_p) of its mass centre.

The output signal from the matrix sensor can be evaluated as two dimensional signal obtained from each pixel U_{mm}. In the case of particle absence it can be determined using following relation [8]:

$$U_{mm} = k \int_{0}^{t_{\exp}} \int_{x_m}^{x_m+\Delta x} \int_{y_m}^{y_m+\Delta y} E(x_1, y_1, t) P(x_1, y_1) dx_1 dy_1 dt, \tag{2.1}$$

where $t_{\exp}$ stands for exposure time, $E(x_1, y_1, t)$ denotes irradiance of point with coordinates (x_1, y_1), $P(x_1, y_1)$ stands for special conversion function that is defined by the quantum efficiency of the photo-sensor and other characteristics, settings of the matrix etc., and k denotes the coefficient of proportionality.

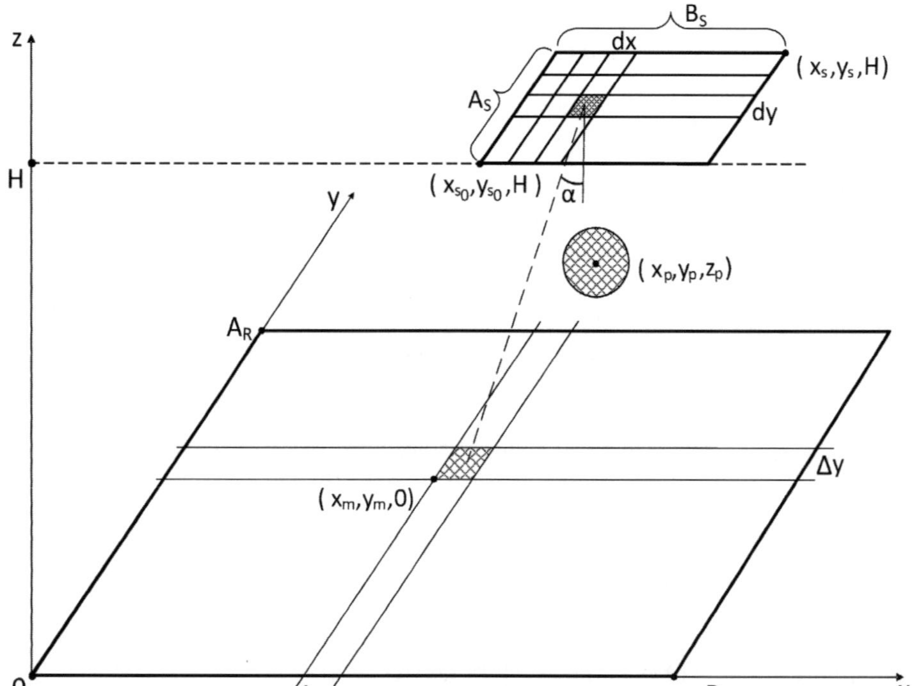

Figure 2. The geometry of the inline CMOS matrix sensor.

The total irradiance of the pixel can be determined in case of constant radiant exitance of light source [8] and absence of particle using following equation [9]:

$$E(x_1, y_1, t) = \int_{x_{s0}}^{x_{s0}+B_s} \int_{y_{s0}}^{y_{s0}+A_s} \frac{M \cdot cos^2(\alpha)}{l^2} dydx, \tag{2.2}$$

where M stands for radiant exitance; α denotes angle between perpendicular line drawn to the surface of the photo-sensor and line that connects the light source unit and the unit of the photo-sensor; l stands for distance between these units. As follows from figure 2, the α angle can be determined geometrically using following equation:

$$\alpha = \arccos \frac{H}{\left[(x-x_1)^2 + (y-y_1)^2 + H^2\right]^{\frac{1}{2}}}, \tag{2.3}$$

If the light unit source illuminates according to the Lambert law [10, 11], the irradiance $E(x_1, y_1, t)$ made by that unit source with radiant exitance M_1 can be estimated as [12]:

$$E_1 = \frac{M_1}{2}\left(\frac{dx_s}{\sqrt{H^2 + dx_s^2}} arctg \frac{dy_s}{\sqrt{H^2 + dx_s^2}} + \frac{dy_s}{\sqrt{H^2 + dy_s^2}} arctg \frac{dx_s}{\sqrt{H^2 + dy_s^2}}\right), \tag{2.4}$$

The total irradiance of the entire matrix sensor can be calculated using integration according to equation (2.2).

If there is a particle between light source and matrix, the relation (2.1) should be rewritten. In order to do so, let us propose that particle absorbs light totally, which means that if the light ray crosses the particle shape, the irradiance made by that light ray is null. In case of not crossing, the irradiance can be estimated using (2.4), therefore, (2.1) can be transformed as follows:

$$U_{mm} = k \int_0^{t_{exp}} \int_{x_m}^{x_m+\Delta x} \int_{y_m}^{y_m+\Delta y} E(x_1, y_1, t) P(x_1, y_l) H(\xi) dx_1 dy_1 dt, \tag{2.5}$$

where $H(\xi)$ is the unit step function [13], ξ denotes the fact of whether the light ray crosses the particles or not. ξ depends on $x, y, x_1, y_1, x_p(t), y_p(t), z_p(t)$. The geometrical task of determining ξ can be considered as the task of crossing between the line and surface (of particle) [14].

The line that connects light source unit (x_S, y_S, H) with matrix unit (x_m, y_m, 0) can be described using following relation [14]:

$$\frac{x - x_s}{x_m - x_s} = \frac{y - y_s}{y_m - y_s} = \frac{z - z_s}{z_m - z_s} \tag{2.6}$$

The (2.6) can be represented in a parametric form with w parameter:

$$\begin{cases} x = (x_m - x_s)w + x_0, \\ y = (y_m - y_s)w + y_0, \\ z = (z_m - z_s)w + z_0, \end{cases} \tag{2.7}$$

In case of crossing between the line and the surface, the following equation should have a real solution:

$$\xi\left[(x_m - x_s)w + x_0 - x_p, (y_m - y_s)w + y_0 - y_p), (z_m - z_s)w + z_0 - z_p\right] = 0 \tag{2.8}$$

The implemented algorithm was based on equations (2.1) – (2.8) and the method of ray tracing. The result is illustrated as a picture, showing the shadow from the particle. The signal without particle (calculated using (2.2) – (2.4)) was also used as a reference in order to determine the equivalent diameter D_e of the particle [3], [15]. The equivalent diameter means the diameter of the sphere with the same volume as a real particle has. The difference between signal with and without the particle can be used for particle size estimation.

The proposed simulation is aimed to determine relation between equivalent diameter of the particle and inline sensor (figure 1) parameters. These parameters are sizes of the light plane A_S and B_S, the distance between this plane and matrix sensor H, the distance between particle and matrix sensor z_p.

3. The results of mathematical modelling

The real sensor dimensions are following: the minimum distance between the particle and matrix sensor is determined by the thickness of matrix's glass (roughly 1 mm) and the thickness of glass plate (roughly 1 mm). The maximum distance measured from the particle to the matrix depends on dimension of the channel (roughly 1 mm), therefore, the particle z_p coordinate varies from 2 to 3 mm. We can present H and z_p in the one figure because they are measured along the same z axis.

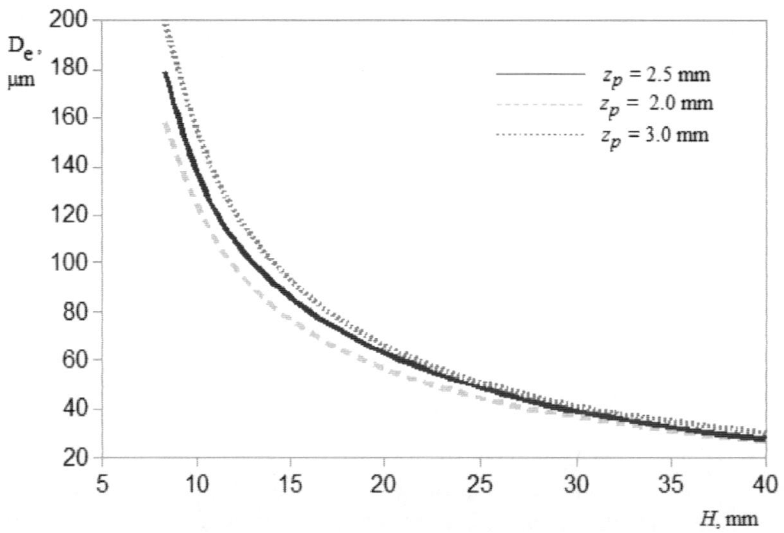

Figure 3. The relationship between equivalent diameter of particle and the distance between matrix sensor and source of light.

Figure 3 presents the relation between equivalent diameter D_e and distance between the matrix and the source of light in case of using 10 μm equivalent diameter of the particle and fixed z_p-coordinates of the particle. The model was calibrated at the H distance of 100 mm.

As follows from figure 3, the equivalent diameter increases significantly depending on the contracted distance between matrix sensor and source of light H from initial size of 10 μm at 100 mm (not shown on figure) to approximately 160-200 μm at the distance 10 mm. The increasing dependence of the z_p-coordinate should also be mentioned.

Figure 4 illustrates the dependence between the equivalent and the real diameter D_r of the particle at the distance 10 mm and 30 mm. The sensor was calibrated for z_p = 2 mm. Figure 4 shows that the particle with diameter 20 μm can be incorrectly measured as particle with diameter ~25 μm in case of H = 10 mm and z_p = 3 mm and roughly the same size if H = 30 mm.

Figure 5 represents the two dimensional discrete signal obtained from the matrix sensor as a result of the simulation using the same conditions as in figure 4.

According to the figure 5 it should be mentioned that the shape of the particles was more easily recognized in the case of 30 mm distance between the matrix and the source of light. The dimensions of the central spot between (c) and (d) vary more than between (a) and (b).

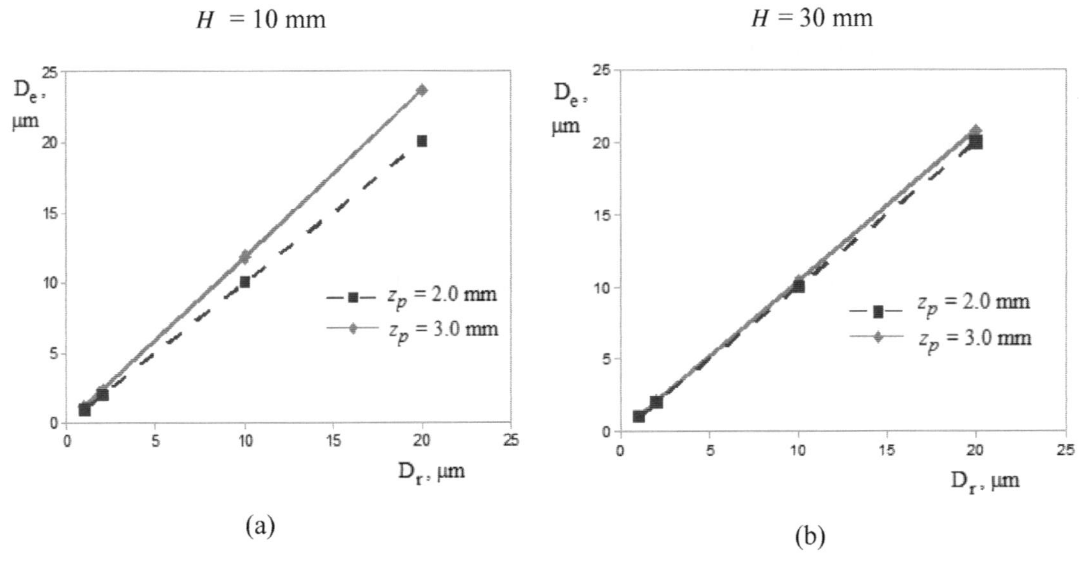

Figure 4. The relationship between the equivalent diameter of the particle and the real diameter. These graphs were obtained for a different distance H between matrix sensor and light source.

IOP Conf. Series: Materials Science and Engineering **153** (2016) 012007 doi:10.1088/1757-899X/153/1/012007

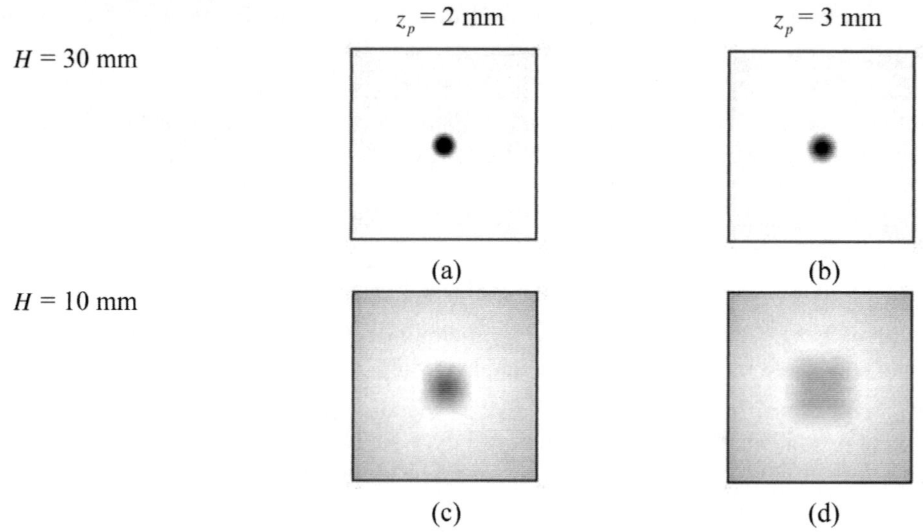

Figure 5. The two dimensional discrete signal obtained from the matrix as a result of simulation.

Figure 6 illustrates the relationship between the equivalent diameter and the area of the light source, S. This area can be determined by the multiplication of A_S and B_S . As follows from figure 6, the equivalent diameter increases simultaneously with increasing light source area, and can be approximated using the square root function. As follows from Figure 7, the increased area of the light source influences the particle shape.

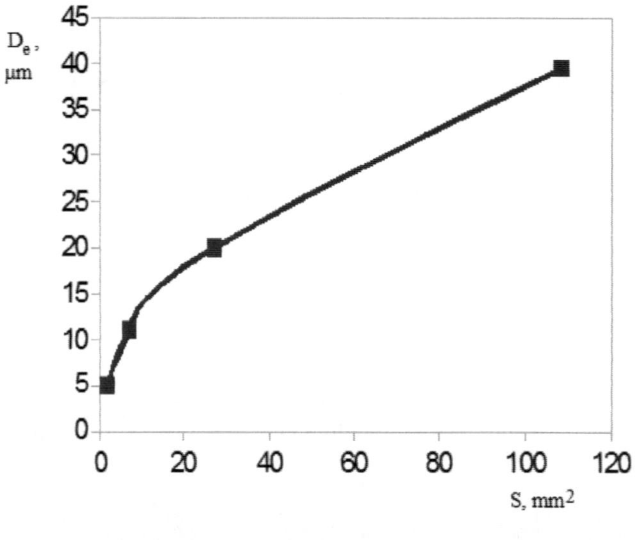

Figure 6. The relationship between the equivalent diameter of particle and the area of the light source.

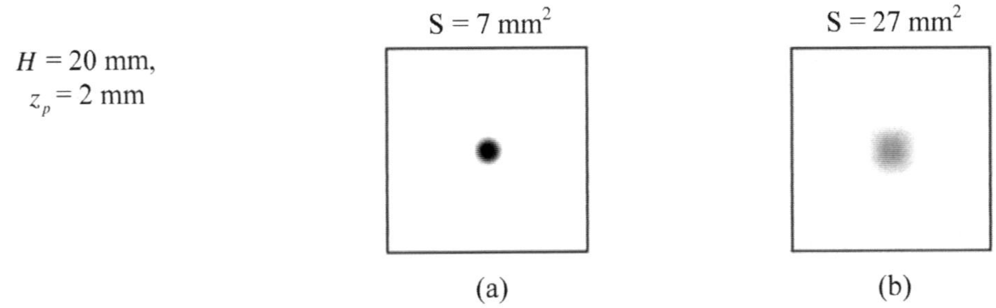

Figure 7. The two dimensional discrete signal obtained from the matrix as a result of simulation.

4. Discussion

As follows from figure 3, the signal level measured as D_e increased significantly as a consequence of decreased distance between light source and the matrix. For example, it was only 10 μm of equivalent diameter at the distance of 100 mm and about 200 μm at the distance of 10 mm. We have here an increase more than 20 times. This dependence might be explained and approved by the inverse-square law. This might be used to improve the sensitivity of the sensor; however, the short distance between the light source and the matrix decreases the accuracy of the sensor. Generally, accuracy is influenced by several factors: one of the most important is the random position of the particle in the sensor channel caused by changing of the flow conditions. The distance between light source and matrix sensor influences the sensitivity of the sensor to the changing of particle z_p coordinate. The differences in z_p coordinate for particles in the sensor channel cannot be lower than approximately 1 mm. Figure 3 shows that particles with the same actual size will be determined as particles with different sizes depending on the distance between the matrix sensor and the source of light. The closer the light source to the matrix, the higher the error in particle size estimation. This is illustrated more clearly on figure 4, where the comparison was made between particle size estimation for two fixed distances 10 mm and 30 mm. The sensor was calibrated at z_p coordinate 2 mm. If the particle flows at the 3 mm z_p coordinate, the measured particle size will be roughly 25 μm instead of the actual size of 20 μm, for the distance H = 10 mm; however, at the distance 30 mm the measured particle size (approximately 20.5 μm) will be almost correct. The difference looks clearer for relatively big particles, therefore, it could be recommended to choose an appropriate distance between the light source and the matrix sensor in order to satisfy both sensitivity and accuracy of the sensor.

The distance H between the light source and the matrix influences the apparent particle shape. As follows from figure 6, for the H distance 10 mm, the shapes appear to be smoother in comparison with those for the distance 30 mm. For the case when z_p coordinate varies from 2 mm to 3 mm, the shadow size changes more in case of H = 10 mm.

An additional parameter that can be changed in order to improve the sensor sensitivity, is the light source area S. As shown on figure 6, the equivalent diameter depends as the square root of the area of the light source. This can also be predictable because the irradiance increases proportionally with the area of the light source, and the signal rises proportionally with the area of the particle. The equivalent diameter should be related to the light source area as the square root; however, the increased source area might lead to the same drawbacks as was shown in case of short H distance, for example, the smoothed particle shape. This effect is illustrated on figure 7. In case of S = 7 mm^2 the shape of the particle shadow is more accurate in comparison with S = 27 mm^2. The other parameters of the sensor was fixed H = 20 mm and z_p = 2 mm.

The results obtained during simulation had good correlation with those anticipated by the basic laws of optics. This can be considered as a positive result of testing the proposed model. The results that were obtained using the proposed mathematical model can advise the implementation during the development of a real inline CMOS matrix sensor by providing an estimate of its characteristics and supporting the choice of the right parameters. First of all, using a relatively short distance between light source and the matrix can be proposed, because this approach leads to the increased signal from a particle; however, using shorter distance might cause decrease of sensor accuracy influenced by increased sensitivity to the z_p coordinate of the particle. It also affects the prediction of the shape of the particle. In addition, the increased area of the light source can be suggested in order to improve the sensitivity of the sensor; however, as shown in figure 7, an increased light source area leads to decreased sensor accuracy similar to the case of contracting the distance between light source and matrix.

5. Conclusion

A mathematical model of the in-line sensor based on matrix photo-sensor has been proposed. The proposed model is based on the assumption that the LED and CMOS matrix sensor can be represented as equivalent planes, with the particle between them, without utilizing any optical elements. The simulation results obtained using ray tracing algorithm show that the optimal characteristic of the sensor can be found by choosing the appropriate distance between the LED and CMOS matrix, as well as the size of LED. It was shown that the signal level is controversial to the accuracy of the sensor. The concrete values of sensor parameters should be determined using proposed mathematical model and obtained results.

An additional investigation should be performed for different pixel sizes[16]. The state-of-the-art matrix sensors have the minimum pixel size lower than 1 μm. In this paper the authors simply intended to demonstrate the principles, but did not consider the effect of pixel size. Further research should take into account the limit of pixel size.

The proposed model should be implemented with more accuracy in case of particle sizes lower than several micrometers, because it might be influenced by interaction with the light. This interaction is described generally by Mie theory and can lead to a bigger shadow than predicted by the classical model. In this study we investigate particles bigger than 10 μm, which have insignificant interaction with light. Further research will be focused on investigation of the dependence between particle sizes in range 0.5 to 10 μm and the particle shadow dimension and its optical density when the light wavelength is fixed.

Another issue is the influence of the liquid flow. This study was aimed to investigate static images; however, there are several cases that should be considered for different particle velocities and exposure time. This might lead to the necessity of processing particle tracks.

References

[1] Fitch J 2002 The Usefulness of Particle Counting in Oil Analysis *EPRI/NMAC Lube Notes* 1, pp 5-25

[2] Michael P W, Wanke T S and McCambridge M A 2007 Additive and base oil effects in automatic particle counters *Journal of ASTM International* **4**(4)

[3] Логвинов ЛМ, Данилов АВ 1995 Математическая модель фотоэлектрического датчика контроля технического состояния трибосистем машин и оборудования *Надежность механических систем* **1** 21–4[Logvinov LM, Danilov AV 1995 Mathematical Model of the Photoelectric Sensor for Monitoring the Technical Condition of Machinery and Equipment Tribosystems *The reliability of mechanical systems* **1** 21–4 (in Russian)]

[4] Логвинов ЛМ 1992 *Техническая диагностика жидкостных систем технологического оборудования по параметрам рабочей жидкости,* ЦНТИ "Поиск", Москва [Logvinov LM 1992 *The Diagnostics of Fluid Systems of Equipment Using the Parameters of the Hydraulic Fluid,* CNTI "Poisk", Moscow (in Russian)]

[5] Кудрявцев ИА 1999 *Повышение разрешающей способности и чувствительности фотоэлектрических преобразователей встроенного контроля дисперсной фазы для систем управления*, диссертация, Самара [Kudryavtsev IA 1999 *Increasing the Resolution and Sensitivity of the Built-in Control of the Dispersed Phase of Photoelectric Converters for Control Systems,* Thesis, Samara (in Russian)]

[6] ISO 21501-1:2009(en) *Determination of Particle Size Distribution — Single particle light interaction methods*

[7] Walker JS, Carruthers AE, Orr-Ewing AJ and Reid JP 2013 Measurements of Light Extinction by Single Aerosol Particles *J. Phys. Chem. Lett.* **4** (10) 1748–52

[8] Стемпковский АИ, Шилин ВА 2003 КМОП-фотодиодные СБИС *Электроника: Наука, Технология, Бизнес* **2** 14–20 [Stempkovsky AI, Shilin VA 2003 CMOS Photodiode SBIS *Electronics: Science, Technology, Business* **2** 14–20 (in Russian)]

[9] Коган ЛМ 1983 *Полупроводниковые светоизлучающие диоды* Энергоатомиздат, Москва [Kogan LM 1983 *Semiconductor Light-emitting Diodes* Energoatomisdat, Moscow (in Russian)]

[10] Иванов ВС 2003 *Основы оптической радиометрии* ФИЗМАТЛИТ, Москва [Ivanov VS 2003 *Basics of Optical Radiometry* FIZMATLIT, Moscow (in Russian)]

[11] Королев ФА 1966 *Теоретическая оптика* Высшая школа, Москва [Korolev FA 1966 *Theoretical Optics* Higher School, Moscow (in Russian)]

[12] Эпштейн МИ 1990 *Измерения оптического излучения в электронике* Энергоатомиздат, Москва) [Epstein MI 1990 *Measurements of Optical Radiation in Electronics* Energoatomisdat, Moscow (in Russian)]

[13] Волков ИК, Канатников АН 2002 *Интегральные преобразования и операционное исчисление* Изд-во МГТУ им. Н. Э. Баумана, Москва [Volkov IK, Kanatnikov AN 2002 *Integral Transformations and Operational Calculus* Publishing House of the MSTU. Bauman, Moscow (in Russian)]

[14] Беклемишев ДВ 2005 *Курс аналитической геометрии и линейной алгебры* ФИЗМАТЛИТ, Москва [Beklemishev DV 2005 *Course of Analytical Geometry and Linear Algebra* FIZMATLIT Moscow (in Russian)]

[15] Беляев СП, Никифорова НК, Смирнов ВВ, Щелчков ТН 1981 *Отико-электронные методы изучения аэрозолей* Энергоатомиздат, Москва [Belyaev SP, Nikiforova NK, Smirnov VV and Tshelchkov TN 1981 *Optoelectronic Methods of Studying Aerosols* Energoizdat, Moscow (in Russian)]

[16] Wedd MW 2010 The Influence of Pixel Numbers and Image Centring upon Particle Size Analysis Accuracy *Part. Part. Syst. Charact.* **27** 71–75

IOP Conf. Series: Materials Science and Engineering **153** (2016) 012008 doi:10.1088/1757-899X/153/1/012008

Through-life data management for composite material products

N Swindells

Ferroday Ltd., Mere Farm Road, Birkenhead, CH43 9TT, UK

E-mail: norman.swindells@ferroday.co.uk

Abstract. The management of digital engineering information throughout the life cycle of an engineered product can be achieved by the representation of this information in information models specified in International Standards. The fundamental basis of this technology is described and its application to the design and manufacture of composites is outlined. The use of ISO 10303-235 'Engineering properties and materials information' for the representation of data from the testing of composite coupons and for the ultrasonic non-destructive evaluation of defects is described as an new example of this technology. The benefits of these standards for the quality control of the information and their role in its conservation are briefly described.

1. Introduction

Throughout the supply chain and the life cycle of an engineered product there is the need for the transfer of digital information about the product between different software systems, each with its own internal method of representing the data. Figure 1 is an attempt to illustrate this life cycle. Digital interoperability between different systems therefore requires a method to ensure that the data in the message from one system can be understood and used by the receiving system. For the long life times of products, such as in the aerospace sector, the digital data also needs to be conserved and yet be still understandable and usable when the originating software is obsolete or no longer available.

The management of engineering information throughout the life cycle of a composite product therefore requires cooperation between several different software systems each with a different role and operation. In an analogy of cooperation between humans from different nationalities, such cooperation requires either pairs of translators between each of the different languages or an agreement to use a common language. The number of translators needed for n individual systems is $n(n-1)$ and so the number of translators required rapidly becomes unsustainable in large networks of systems. All software systems use a proprietary internal information structure and a local internal dictionary and it would be more efficient and economical if all the members in a communication chain used a common language, based on a common information structure and a common dictionary. The solution adopted by the global engineering profession was therefore to develop a standard information structure for the *external* exchange of the data and a standard structure for a dictionary. This solution effectively provides a common language to promote interoperability and cooperation between different engineering systems.

IOP Conf. Series: Materials Science and Engineering **153** (2016) 012008 doi:10.1088/1757-899X/153/1/012008

A solution to digital interopeability was identified as being necessary because of the global expansions of the supply chains for the main capital industries. These requirements arose almost simultaneously in all the main industrial nations and there has been a corresponding global response with a high degree of collaboration and mutual understanding to resolve the problems that were identified. However, the use of the technology has remained a preserve of a few companies and industrial sectors and there has been limited awareness and adoption by a wider population. A recent report from the Strategic Standardisation Group of the Association of Aerospace and Defence Manufacturers of Europe (SSG-ASD) is titled: "*Through Life Interoperability – A critical strategic lever for competitiveness*" [1]. The Group's vision for through life interoperability is that "All players in the aerospace value network will be able to share digital information securely throughout the life of the products and services". The report recognizes the value of standardized information models for product data representation and exchange for this role and intends to make these standards the cornerstone of this interoperability. Some of the standards for the representation of data for composite products on which this strategy depends are described in this paper and an overview of the technology and more information in its applications has been published [2].

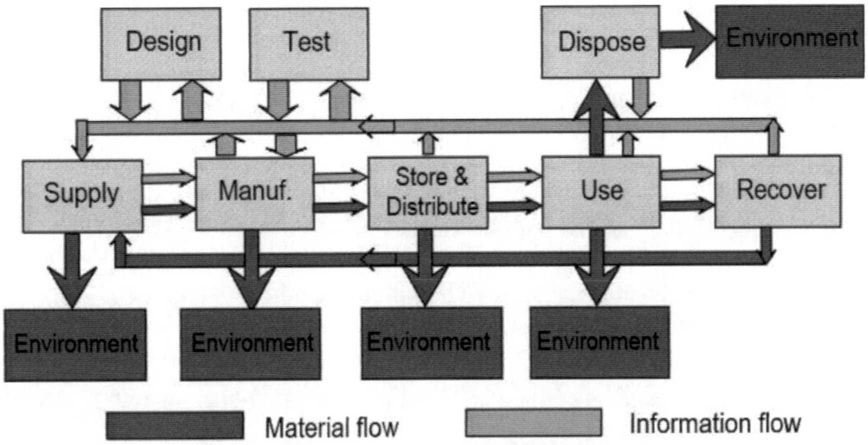

Figure 1 Illustration of the flow of information though a product life-cycle [© Ferroday Ltd 2015, used with permission]

2. Standardised information models

The solution for interoperability and information conservation of product data adopted by the aerospace, automotive, defence, construction, machine tools, process plant, offshore oil & gas and other sectors is to represent the product data in non-proprietary, computer-understandable information models. The models are specified in a series of International Standards to provide a common language for the digital representation of engineering information. These Standards have been developed since 1984 within the ISO Technical Committee 184, Sub-committee 4 (ISO/TC 184/SC4) in a collaborative effort by hundreds of engineers from the main industrial nations and most of the global industrial sectors. The models are written in the EXPRESS computer processable language [3] standardised in ISO 10303-11 and look like object oriented computer algorithms.

Information models in the ISO 10303 family of standards are of two types:

- Integrated Generic Resources (IGR) – a representation of the basic concepts of engineering and manufacturing in a single generic information model;

- Application Protocols (AP) – selections from and extensions of the generic model to represent specific industrial situations. It is an AP that is implemented in, or interfaced to, engineering application software for real-time processing, communication and archiving of the data.

The information models are developed by very rigorous, formal methods to describe, in a computer processable form, aspects of the real world of manufactured products, manufacturing processes and their properties. In effect, they provide engineering specifications for this data with the important benefit that they are independent of proprietary software. These specifications can also be used as the basis for quality control and quality assurance of the engineering data, just as with any other engineered product. The additional value of these information models is that they conserve both the syntax and the semantics of the data, ensuring that the information that the data represents can still be understood and interpreted correctly in the future even if the originating software is no longer available. They have been used for the representation of engineering data for over 20 years and the reliability of this representation has been proven many times.

Some aspects of the scope and application of some of the standards that are relevant to the life cycle of composite materials products are summarized in the following sections.

3. Design and manufacture of composite material products

Hunten [4, 5] has described the application of ISO 10303 standards to the design and manufacture of composite material products. The problem of integration of information between shape design, analysis and manufacture is made more difficult when the product is made from a composite material structure. Most production systems currently employ specific point-to-point translators to enable this transfer of information, and very few analysis systems are able to return geometric shape information seamlessly to the design shape modeler. With composite structures there are the additional problems of calculating true fiber directions and ply flat patterns that are shared with the manufacturing process. All of this information needs to be shared with commercial or in-house detailed analysis codes such as those for fastened joints and panel buckling. The interconnections required between design, analysis and manufacturing information for a composite product are illustrated in Figure 2.

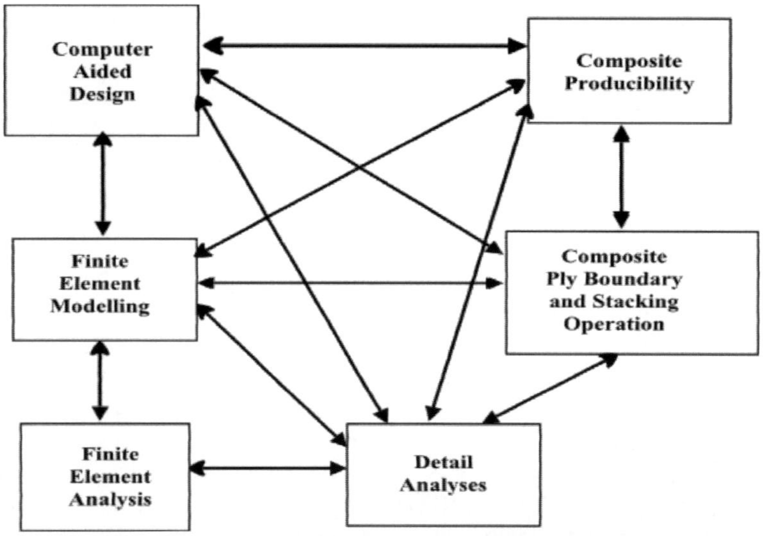

Figure 2. The interconnection of design, analysis and manufacturing information [© Hunten 2014, Reproduced with permission]

IOP Conf. Series: Materials Science and Engineering **153** (2016) 012008 doi:10.1088/1757-899X/153/1/012008

Edition 1 of ISO 10303-209:2001 Composite and metallic structural analysis and related design (AP209) was developed in ISO TC184/SC4 to enable companies using different CAD and FEA systems to exchange engineering design and analysis data using standard file formats. Configuration management data within the standard can ensure that design and analysis information carried out by the different teams relate to the correct product versions. A company will be able to archive associated, configuration managed CAD and FEA data in the AP209-format with confidence that the data can be reused in the future whatever changes of systems have occurred.

The information model of Edition 1 of AP209 provided resources to describe the following main concepts:

Finite element data: This includes models, analysis definitions and load cases, and results. A model can be specified in as much detail as required – if necessary down to the level of element shape functions, discretisation points and integration rules. Static and natural frequency analyses are within the initial scope.

Configuration management data: A version of the finite element model is linked to a version of the product. This ensures that the correct finite element data may be associated with the correct version of a product within a Product Data Management (PDM) system.

Product geometry: Both the design geometry and the idealised geometry created for analysis can be recorded. Nodes, finite elements, their edges, faces and volumes can be explicitly associated with aspects of the product geometry. It is possible to specify element properties, loadings or boundary conditions on a curve, edge, surface, or volume of the geometric model.

Composite lay-up: The lay-up of a composite part can be specified in detail. Shape, stacking sequence, and property information can be supplied for individual plies and their fibre orientations.

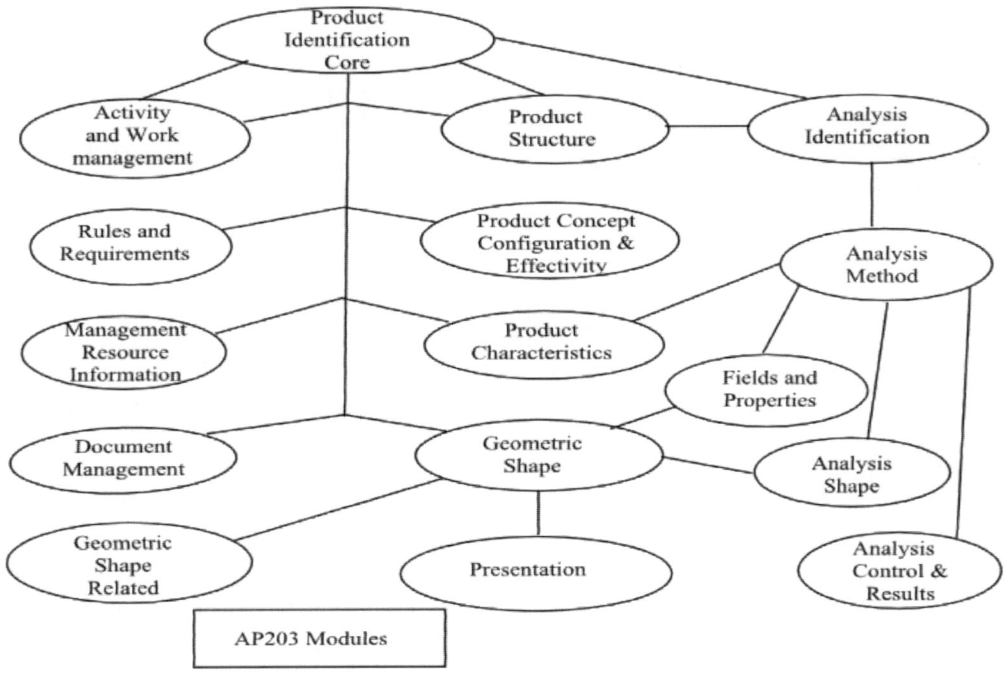

Figure 3. A high level overview of ISO 10303-209 edition 2
[© Hunten 2014, Reproduced with permission]

IOP Conf. Series: Materials Science and Engineering **153** (2016) 012008 doi:10.1088/1757-899X/153/1/012008

The second Edition of ISO 10303-209:2014 (AP209 ed2) – now renamed Multidisciplinary analysis and design [6], has added a generic Engineering Analysis capability complimented by specific Computational Fluid Dynamics (CFD) and a generalized mesh based numerical analysis capabilities to the classical Finite Element Analysis capabilities of AP209 Edition 1. The intention is that AP209 ed2 will now address a much wider set of multi-disciplinary analysis and optimization problems. A high level overview of AP209 Edition 2 is shown in Figure 3.

Edition 2 of AP209 has also been incorporated into the new development of ISO 10303-242 Model-based 3-D engineering.

4. Engineering properties and materials information

Ferroday Ltd is responsible for the development of two information models in the ISO 10303 series. ISO 10303-45: *"Material and other engineering properties"* is a part of the IGR and is a generic information model for the representation for any property of any product and its value, including the chemical composition [7]. ISO 10303-235: *"Engineering properties for product design and verification"* is an AP that extends the IGR to represent the collection of processes by which the value of any property of a product is obtained [8]. The scope of ISO 10303-235 is summarised in Table 1.

Table 1. Summary of the scope of ISO 10303-235

Technical	Administrative
Products, product history	Persons, organizations, addresses
Properties of products and resources	Approvals, qualifications, certifications
Processes and their properties	Dates, times
Numerical and descriptive values	Documents and files
Mathematical expression values	External references
Uncertainty and reliability of all values	Effectivity
Product substance composition and structure	Language
Dimensional and positional tolerances	Locations
Resources	Requirements

The principal concepts on which ISO 10303-45 and ISO 10303-235 are based are:

- All 'materials' are products produced by a process and so can be represented by the information models for products and processes developed by ISO/TC 184/SC4.
- The process used to determine the value of an engineering property of a product defines its meaning (semantic).
- The magnitude of the value of an engineering property will be dependent on the characteristics and operation of the measurement method and on the characteristics of the product.

As examples: most property values vary with temperature, some property values depend on the dimensions of the product; the value of the tensile strength of a metal depends on the rate of the extension of the load applied during the test. The collection of characteristics and their values that have an influence on the value of an engineering property are represented by the data environment, defined in ISO 10303-45, and these need to be associated with the property value in order to conserve the semantics of this value.

There are many processes that are used to determine the engineering properties of a product and they vary between different national industrial systems. It is therefore not practical to include the names and definitions of measurement processes and the names of their properties in the information model of ISO 10303-235. ISO 13584 *Parts libraries* defines an information model to provide a dictionary of terms and their definitions and ISO 10303-235 can make a reference to this external source to obtain these names. Kafka [9] has shown that a dictionary of testing processes and their related properties can be developed from these principles.

It was discovered in the development of ISO 10303-235 that there is a level of abstraction that enables all measurement processes to be represented by instances of the same information structure. So that, combined with the capability to reference a dictionary for the names of the measurement processes and their properties, ISO 10303-235 can represent any property of any product measured by any method.

5. Applications of ISO 10303-235

5.1 Method

For the representation of data for each of the examples of applications of ISO 10303-235 the procedure followed a series of stages:

1. The starting information would be in the form of a data sheet or other document describing some of the details of the measurement method, the details of the equipment used and the results.
2. The document form of the information was represented by a high-level information model using EXPRESS-G, the graphical form of EXPRESS, in order to ensure that all of the information in the starting document was accounted for and could be understood.
3. A *view,* or portion, of the standard model was then created by software to form a template for the representation of the information in the high level model. The relationship between the data items in high-level model and the items in the view of the standard model was confirmed by documenting the mapping relationship between them. The template view of the standard model can then be managed by the TenDEX® software developed by Ferroday Ltd to enable the input of the data for the application.
4. The TenDEX® software can be configured to present a data sheet for the input of data into the standard structure that corresponds to the real world example in stage 1. Inputting data into the template via the software automatically creates a data file as an instance of the template model. The format of this data file conforms to ISO 10303-21 and conserves the information structure of the standard model as well as the data values in those parts of the model that were selected to represented the real world results. The template can be reused each time the test or investigation is carried out.

5.2 The testing of composite coupons

The testing of coupons extracted from sheets of prepared composite materials is a starting point for the life cycle of a composite product. The preparation of these sheets for aerospace materials and the manufacture of the coupons from these sheets is governed by European and aerospace standards. Following the method outlined in 5.1, above, a selection of relevant sections of ISO 10303-235, forming a view of the standard as a template, was able to represent the testing of five coupons manufactured in this way for both tensile strength and buckling by compression between two grips. This view of ISO 10303-235 was able to record the history of the process of manufacture of the test specimens by reference to the appropriate standards, the characteristics of the testing machine and the conditions of its operation, the values of the properties and the statistical evaluation of the results.

5.3 Ultrasonic non-destructive evaluation (NDE)

The results of a non-destructive evaluation of a product for internal defects depend on the settings of the control instrument, the characteristics of the ultrasonic probe and the capabilities and level of competence of the operator. Interpretation of the engineering structural integrity of the product requires a detailed description of the product and there may be more than one product joined together in the structure being tested. Applications of ISO 10303-235 were created, following the method outlined in 5.1, above, to represent all of the details of three products and their ultrasonic NDE inspection by the methods of phased array, pulsed echo and time-of-flight diffractometry (ToFD) respectively. Views of ISO 10303-235 were developed for each of these use-cases to provide templates for the input of the characteristics of the product, the control instrument and the probe and a record of the locations of the defects that were detected. The qualifications and certification of the operator could be represented as well as the qualification and certification of the supervisor who specified the method to be used and the details of the verification of the process and the results.

6. Quality control and quality assurance of product information

With the change to all digital representations of engineering information there is a critical need to ensure that the data that represents the information can be relied upon, that it is supplied to agreed specifications and meets the needs of the users. These are the concepts of quality assurance of any product, which should be supported by formal methods of quality control of the information by the supplier – just as with any other engineered product. They are of greater importance when the data has to be conserved for the whole life cycle of a product. The requirements for the quality control and quality assurance of information and data are being specified with the development of ISO 8000 – Information and data quality. ISO 8000 Part 8: *Concepts and measuring* [10] provides:

- Definitions of the quality of information and data;
- A structured way to plan measurements of the quality of information and data;
- Pre-requisites for measuring the quality of information and data;
- Requirements for reporting measurements of the quality of information and data.

Information is defined by ISO 8000-8 as: "*Knowledge concerning objects, such as facts, events, things, processes or ideas, including concepts, which has particular meaning within a certain context.*" Data is defined as: "*Re-interpretable representation of information in a formalized manner suitable for communication, interpretation or processing.*" *Metadata* is data that describes and defines other data. *A conceptual model* is the model that describes concepts of a universe of discourse.

Three types of quality are defined for information and data: *syntactic*, *semantic* and *pragmatic*. These concepts are intended to apply to any information and data. Swindells [8] has shown how the standards developed in ISO/TC 184/SC4 conform to the recommendations of ISO 8000-8.

7. The conservation of digital data

The conservation of digital data in an archive will be necessary to manage the information from the composite product throughout its life. This requirement is common in all sectors where the information is now mostly created and processed digitally. This common requirement can be managed by a procedure originated by NASA for its space data and now adopted by ISO and standardized as ISO 14721 [12]. The procedure is known as the Open Archival Information System (OAIS framework) and has six stages:

1. *Ingest* – input data records plus the metadata to enable them to be found in the future;
2. *Data management* – ensure quality control and quality assurance of the input;
   ISO 10303 specifications can provide the basis for quality control;
3. *Storage* – physical storage of data including back-up and migration to new media;
4. *Preservation planning* – ensuring the data is not just a stream of unreadable digital bits when future access is required and the original software and systems are no longer available;
   ISO 10303 standards conserve the semantics as well as the syntax of the data;
5. *Access* – finding records in the archive and distributing them to appropriate users;
6. *Administration* – operation and maintenance of the archive.

Note that the second and fourth stages will benefit from the use of the standard information models from ISO/TC 184/SC4.

8. Discussion

The main aims of this research were to demonstrate that the global knowledge base provided by the information models in the ISO standards for data representation and exchange can be applied to the representation of real engineering information for composite material products and also to develop ways to reduce the barriers to their use. The standards from ISO/TC 184/SC4 do not provide any indication of how they may be used as an application for an industrial information representation that is within the scope of the standard. The technology of these standards is at the interface between deep knowledge of an industrial engineering situation and the understanding of modern software and information modelling. Very few engineers possess this combination of knowledge and experience and most would not have the time, the opportunity or the wish to acquire it, however, they would want to be able to verify that the computer representation of what could be critical information is accurate and relevant for their needs. The method described in this paper of generating a view of ISO 10303-235 as a template for the input of data for a particular application provides a way of validating that the standard model can represent a real application. The high level-model representation of the original data and the documentation of the mapping onto the template derived from the standard model could be sufficiently concise and clear to enable users to review their scopes and confirm that the template meets their needs. A template derived in this manner could be re-used many times for the input of data from a series of the measurement processes that it represents.

ISO 10303-209, and its incorporation into ISO 10303-242, is intended for use at the design stage of the composite material product. The subsequent stages of the life of the product will be monitored by successive measurement of the properties of the product that are appropriate for the stage of life. Since ISO 10303-235 can be used for the representation of the combination of product, process and property data, this standard can be used for recording the status of all of the subsequent stages in the life including monitoring the structural integrity by NDE. Applying the principles established in ISO 8000-8 should ensure that this series of property data records could be relied upon to provide a history of the product and an audit trail for the source of engineering information for the whole life cycle of the product. The OASIS framework provides the means to ensure that the data will still be available for use in the future.

9. Conclusions

The digital representation of engineering product, process and property information based on the global knowledge in the standards from ISO TC184/SC4 is:

- *Comprehensive* – applicable to all engineering activities;
- *Expandable* – able to cope with new types of products and processes;
- *Scalable* – able to represent small and large data sets;
- *Sustainable* – independent from proprietary software;
- *Dynamic* – in continuous development to match new developments in IT methods and resources;
- Supports improved interoperability and quality assurance to reduce costs.

The application of this technology to the representation of data for composite products has been proven with both ISO 10303-209 and ISO 10303-235 and these standards should be adopted more generally by this industry sector.

References

[1] Aerospace and Defence Industries Association of Europe – Strategic Standardisation Group 2014, *Through Life Cycle Interoperability* http://www.asd-ssg.org/through-life-cycle-interoperability, accessed 2016-02-16.
[2] Moreno A 2014 *Interoperability for digital engineering systems* (Milan: FrancoAngeli)
[3] Schenk D and Wilson P 1994 *Information modelling the EXPRESS way* (New York: Oxford University Press)

[4] Hunten K A 2014 Design and manufacture of composite products *Interoperability for digital engineering systems* ed A Moreno (Milan: FrancoAngeli) pp 61-66

[5] Hunten K A 2014 ISO 10303-209: Composite material structure, shape and properties *Interoperability for digital engineering systems* ed A Moreno (Milan: FrancoAngeli) pp 118-129

[6] ISO 10303-209:2104, *Product data representation and exchange – Part 209 : Application protocol: Multidisciplinary analysis and design*, (Geneva: International Organization for Standards)

[7] Swindells N 2000 Communicating materials information – product data technology for materials *Int. Materials Reviews* **47**(1) 31–45

[8] Swindells N 2009 The representation and exchange of material and other engineering properties *CODATA Data Science Journal* **8** 190–200

[9] Kafka V 2003 Reference dictionary for product properties *Metal fabrication and welding technologies* ed A A Becker (Loughborough: Q3 Digital) pp 273–281

[10] ISO 10303-8:2016, *Information and data quality – Part 8: Concepts and measuring* (Geneva: International Organization for Standards)

[11] Swindells N 2014 ISO 8000-8: The quality of information and data *Interoperability for digital engineering systems* ed A Moreno (Milan: FrancoAngeli) pp 202-206.

[12] ISO 14721:2012 *Space data and information transfer systems -- Open archival information system (OAIS) -- Reference model* (Geneva: International Organization for Standards)

Design and research of thermal protective material from short basalt fibres

MA Komkov, VA Tarasov, RA Boyarskaya and AS Filimonov

BMSTU, 5, 2nd Baumanskaya St., Moscow, 105005, Russia

alexf72@mail.ru

Abstract. Design and manufacture issues regarding highly porous thermal protection coatings of products by means of liquid filtration of short basalt fibres and mineral binder are considered. The technological process of manufacture of thermally loaded products from the short basalt fibres of thermal protective material (TPM) in the form of tiles and rings, was developed based on a liquid filtration method. The structural and mechanical properties of the highly porous TPM technological modes were determined. The thermal testing of the pipe model samples was carried out on a thermal bench, which showed the temperature on the coating reaching less than 60°C during a hot air run through the pipe at 400°C.

1. Introduction

Rocket and spacecraft structures are under considerable thermal load both during operation in orbit, and passage through dense atmospheric layers. To prevent rocket and spacecraft structures from overheating, the complex system of thermal protection, consisting of 4 layers, is used: a solid layer of metal or composite materials; a layer of material executing heat insulation at the expense of low heat conduction; a layer of char-forming material in which endothermal reactions take place; and a layer of material which is carried away with incoming flow and creates radiative energy flow from rocket and spacecraft surfaces. Therefore, the creation of highly porous, low-density, and environmentally friendly thermal insulation from available and low-cost basalt fibres for rockets and spacecraft operation at temperatures up to 700°C, is an important scientific and engineering task [1, 2].

As a material for heat insulation coating of rockets and spacecraft items operating for a long time at temperatures up to 400°C and more, the thermal-protective material of TZMK-10 grade can be used [1]. The material is based on fine quartz fibre with a diameter of 1.5 – 2.0 μm, and was developed for the heat protection tiles for the spacecraft "Buran". The disadvantages of the material TZMK-10 are the high cost of its production and the high coefficient of thermal conductivity (0.06 W/(mK)) at normal temperature.

A cheaper but no less effective heat insulation can be produced with short basalt fibre. This thermal-protective material (TPM) is comprised of superthin (diameter 0.5 – 3.0 μm) basalt fibres (BSTF) with alumina binding material, and its heat insulation is based on glass microspheres of МС-ВП-А9 [MS-VP-A9] grade and organosilicate compositions ОС-11-07 [OS-11-07]. It was deduced from experiments [3 – 5] that when heated to temperatures of 400-500°C, this material would have a low coefficient of thermal conduction, in the mean equal to 0.041-0.042 W/(mK).

Spheroplastics have double the density in comparison with the heat insulation made of the short basalt fibres with mineral binder, but at high temperature and over long-term heating, the spheroplastics are charred and cracked. In this case, TPM based on basalt fibre, demonstrates high cycle fatigue strength, maintains its shape quite well, and its chemical stability is significantly higher in comparison with that of heat insulation made from spheroplastics. The main method of manufacture of short basalt fibres based TPM is a method of fibre filtration deposition from liquid suspension (pulp), which is used in production with thermal protective tiles and short cylindrical rings [2, 6].

The objective of this paper is to determine parameters for the manufacture process of high porosity, thermal insulating coatings of spacecraft structures, from short basalt fibres and mineral binder, by filtration deposition.

2. Filtration analysis

The technology for the production of short-fibre heat-insulating coatings involves a combination of individual operations including cleaning and fragmentation of the original fibres, preparation of hydro-mass (pulp) from the short fibres, and liquid formation and thermal processing of the product. In this case, the process of filtration deposition of fibres onto a production rig determines the coating material structure and, consequently, its thermo-physical and mechanical characteristics.

The filtration rate depends on the hydro-mass properties – the temperature and operating fluid viscosity, the sizes of the fibres, and the shape of the solid phase particle – binder. The filtration conditions are determined by the hydrostatical pressure value, the filtrate chamber vacuum treatment, and the cake layer porosity value. Utilization of low pressures for pulp with low fibre content improves the structural, thermal and physical characteristics of the moulded product, at the expense of the fibre distribution over material volume.

The mathematical modelling of the moulding process of the short-fibre material heat-insulating structure should be based on the physical principle of liquid flow through a porous body. In this case, the model constants, dimensional or non-dimensional parameters should be used. They are easily determined experimentally and are dependent on the physical properties of the liquid and structural characteristics of fibrous material.

The liquid moulding process for the manufacture of rocket and spacecraft pipes comprised of short basalt fibre heat insulation material is time consuming. The determination, expressed in analytical form, of factors affecting the filtration rate and their inter-dependence, is an important task for the development of the manufacturing process of products based on the short-fibre compositions.

The basis of filtration physico-mathematical analysis is a description of the liquid flow (filtrate) into the pores between fibre layers. The true speed of fluid flow into the pores between fibre layers and binder particles, is difficult to determine since it is impossible to take into account the size and configuration of each pore. Therefore, instead of the true speed of the liquid flow, its conditional value, called the filtration rate, is used.

The filtration rate, or fluid flow, $u_f(t)$, through the porous layer of the already deposited fibres (figure 1), is determined by Darcy's law [7]

$$u_f(t) = \frac{K_p p}{\eta_{\text{liq}} h_{\text{layer}}}, \quad (1)$$

where K_p (m^2) is the coefficient of liquid permeability through a porous medium; p (Pa) is the pressure differential on the fibre layer; η_{liq} (Pa·s) is the dynamic viscosity of the filtered fluid; h_{layer} (m) is the thickness of the fibre sediment layer.

In equation (1), if we substitute the pressure, p, for the sum of the liquid column height and vacuum pressure Δp of the filtrate chamber, and the current set of the sediment layer thickness via liquid volume filtered through the perforated partition $h_{\text{layer}} = c_w(H - h)/\rho_{\text{layer}}$, then we obtain a differential equation that determines the filtration time of the pulp:

$$u_f(t) = \frac{K_p(\rho_p g h + \Delta p)\rho_{\text{layer}}}{\eta_p c_w (H - h)} = \frac{dh}{dt}. \quad (2)$$

Here ρ_p (kg/m^3) is the density of the fluid (pulp); g (m/s^2) is the acceleration due to gravity; h (m) is the current height of the liquid column; ρ_{layer} (kg/m^3) is the density of the dry sediment, for the average layer thickness; c_w (kg/m^3) is the weight content of fibre in the pulp; and H (m) is the initial height of the liquid column (hydro mass, suspension, pulp).

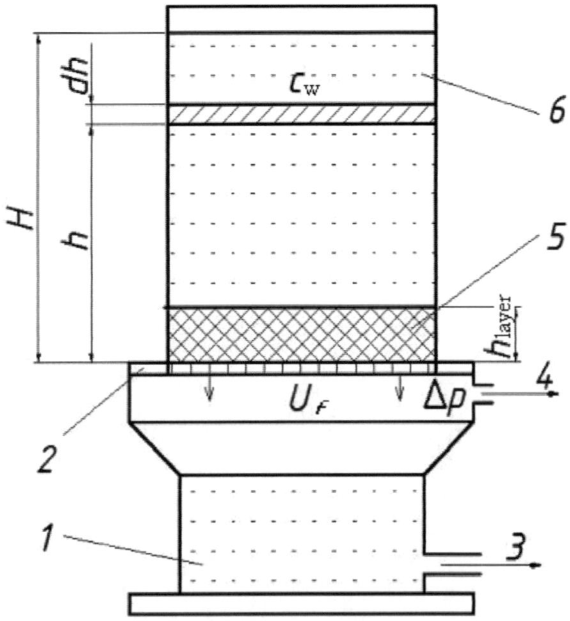

Figure 1. Schematic showing the filtration deposition of short basalt fibres of pulp onto a perforated mandrel: 1 - filtrate; 2 – perforated partition; 3 – filtrate drain; 4 – chamber vacuum treatment; 5 – fibrous sediment layer; 6 – pulp

After the integration of equation (2) we investigate the time required for the deposition of fibres from the pulp according to the scheme shown in Figure 1:

$$t_f = \frac{\eta_p h_{\text{layer}}}{K_p \rho_p g}\left[\frac{h_{\text{layer}}}{H} + (1 + \Delta p_H) ln\left(\frac{1 + \Delta p_H}{h_{\text{layer}}/H + \Delta p_H}\right) - 1\right]. \quad (3)$$

In equation (3) $\Delta p_H = \Delta p / H \rho_p g$.

If the filtrate chamber vacuum treatment is neglected, *i.e.* Δp=0, then from (3) one can conclude

$$t_f = \frac{\eta_p h_{\text{layer}}}{K_p \rho_p g}\left[\frac{h_{\text{layer}}}{H} + ln\frac{H}{h_{\text{layer}}} - 1\right]. \quad (4)$$

According to the filtration equation the pulp volume is determined and the productivity of the product moulding process of the designed manufacturing equipment is calculated. The majority of the filtration time corresponds to operation at lower concentrations of pulp and a lower hydraulic pressure.

In equations (1-4) the coefficient of permeability K_p is the permeability relative to fluid through a porous medium under the action of an applied pressure gradient, it has the dimensions of area, and a value of the order of the square of the size of a typical pore. The coefficient of permeability is a structural feature by its nature, and is determined mainly by the geometry of the pore space. According to Kozeny [8], it depends on the specific volume of the open pores $n = V_p/V_{\text{layer}}$ cubed, divided by the square of the fibre specific surface s_y, slowly streamlined by liquid:

$$K_p = \frac{cn^3}{s_y^2} = \frac{cn^3}{s_0^2 v_f} = \frac{cd_f^2 n^3}{16(1-n)}. \tag{5}$$

Here V_p (m^3) is the volume of pores in the sediment layer; c is a dimensionless quantity: for cylindrical pores $c = 0.5$ and for pyramidal pores $c = 0.597$; $n = 1 - p_{\text{layer}}/p_f$ defines the porosity of the fibrous media; p_f (kg/m^3) is the density of the fibre; s_0 (m^2) is the specific surface of the fibre unit length, $s_0 = s_f/V_f = 4/d_f$; d_f (m) is the average (effective) diameter of the fibres; and $v_f = 1 - n$ is the fibre volume content in the sediment layer. In equation (5), recorded later by Karman [9], the value of $(1 - n)$ was taken in the second degree. Although squared, in our opinion, it should only take account of the specific surface of the fibre s_0, but not their volume fraction in a porous material. We have observed that, when a large porosity, for example, $n = 0.9K_p$, is calculated according to the Karman formula (equation 5) and determined experimentally, there is a difference of almost an order of magnitude. For porosity of $n = 0.30 - 0.35$, specific to structural TPM, these differences are not more than 30% and with some variations of fibre diameters in composite material, they become hard-to-find in K_p calculations.

To assess the applicability of formula (5), we have conducted experimental studies to determine the permeability coefficient of the fibrous samples, using service water on a special plant. TPM model samples of $A = 100$ (cm^2) of the same surface density (weight) $\gamma^* = p_f(1 - n)h_{layer} = 0.60$ (g/cm^2) and made of quartz zero-twist yarns with a linear density of 104 (mg/m) and where the in yarn fibre diameter is 12 microns. Short fibres were produced by cutting yarn in a special device into segments with a length of 6.7 mm, 13.4 mm and 20.1 mm. The quartz fibre density was $p_f = 2.13$ (g/cm^3).

During liquid impregnation of the fibrous samples, the pressure difference across the layer Δp was measured by vacuum gauge. The liquid volumetric flow rate Q was measured by flow meter and layer thickness h_{layer} was also measured, and the porosity of the sample was calculated according to formula

$$n = 1 - M_{\text{sample}}/(Ap_f h_{\text{layer}}), \tag{6}$$

where $M_{sample} = A\gamma^*$ is the sample mass. According to the water flow experimental results, the experimental values of the permeability coefficient were calculated by the equation (1), and can be written as:

$$K_p = \frac{Q\eta_{\text{liq}}}{A(\Delta p/h_{\text{layer}})}. \tag{7}$$

3. Experiments

Figure 2 presents the analytic dependence of the permeability coefficient of the porous fibrous material, calculated using formula (5), and experimental values of the porosity n and the coefficient K_p, calculated using formulas (6) and (7). Figure 2 demonstrates that the dependence calculated according to equation (5) is reasonably consistent with the results of the fibrous samples obtained from the water flow experiment. In figure 3, the experimental dependence of samples porosity of the cut fibre length is presented. From figure 3, it is shown that the shorter the fibres, the greater material porosity.

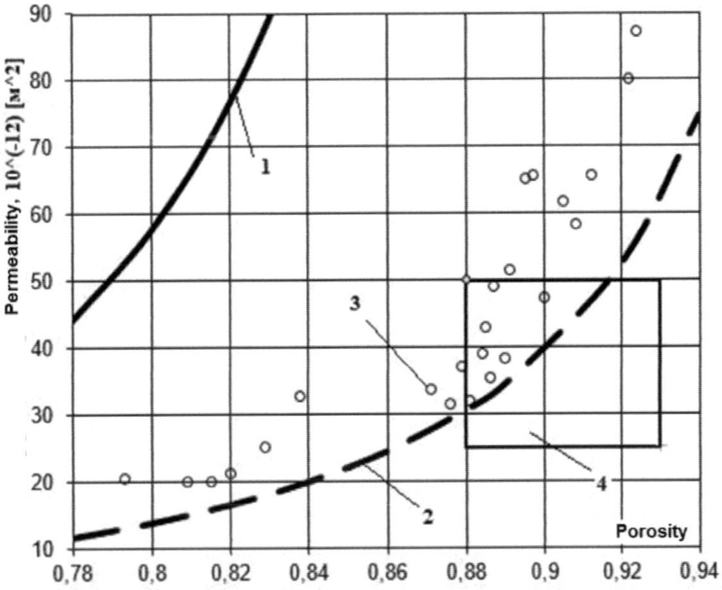

Figure 2. The dependence of the permeability coefficient K_p of the cut fibre samples on the material porosity n: 1 – by the formula of Kozeny-Karman; 2 – by formula (5); 3 – experimental values; 4 – the results area of experimental studies on glass fibre samples [8]

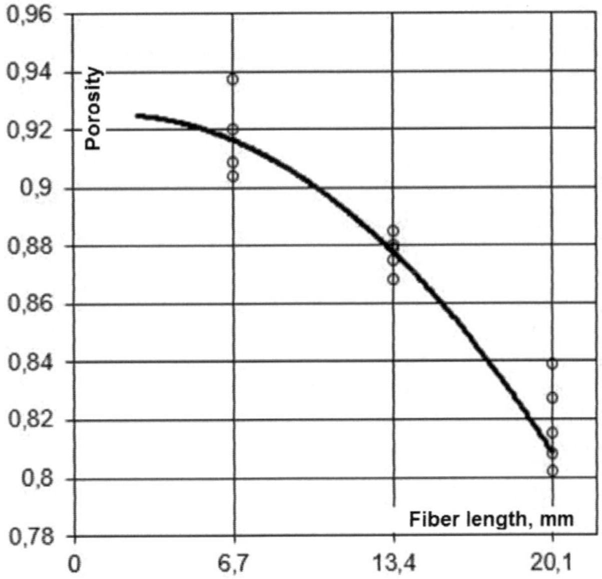

Figure 3. The dependence of the sample porosity on the length of the short fibre

At the same time the fibre length cannot be too small, less than $l_f = 1.0 - 1.5$ mm, because during the filtration deposition of very short fibres they can partially lie down at an angle or perpendicular to the sample or product plane. This reduces the strength of the composition for compression, and the material porous structure becomes unstable. On the other hand, with fibres of decreasing length (figure 3) the TPM porosity increases, thereby reducing the material thermal conductivity of the material. To obtain a fibre length of 1.0 – 1.5 mm, a further shortening of the original cut fibres is carried out in propeller stirrer. The relationship between fibre length and processing time is shown in figure 4. When the concentration of fibres in water is $c_w = 15 - 20$ (g/l) and the rotation speed of mechanical stirrer is 1200-1500 rpm, the disintegration time of the quartz fibres and basalt fibres is 3 – 5 minutes and 8 – 15 minutes, respectively.

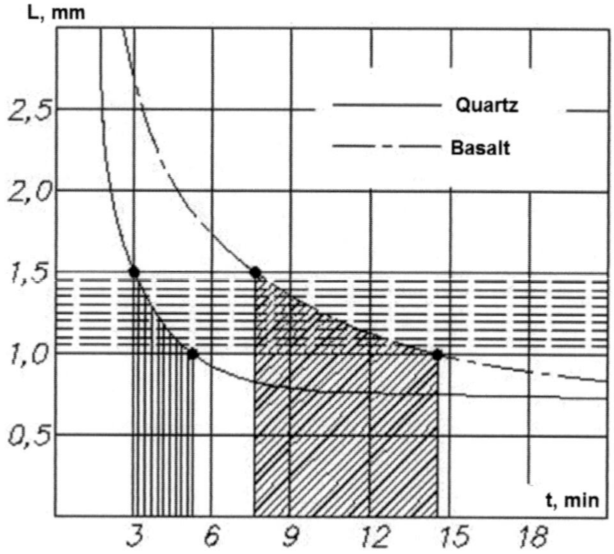

Figure 4. The relationship between fibre length and processing time:
1 – silica fibre; 2 – basalt fibre

The choice of the minimum possible diameter of fibre for the basalt TPM manufacturing is constrained by the fact that for a diameter of less than 3 μm [10] the thermal conductivity becomes too low (figure 5). Besides, with small diameters, the thermal resistance of the material increases at the expense of the specific surface increase; however, when the fibre diameter is less than $d_f \leq 1.5 - 2.0$ μm, and when the wavelength of the thermal spectrum becomes comparable to its diameter, the electromagnetic wave can bend around the fibre. As a result, the composition becomes transparent for thermal radiation and consequently the fibrous composition thermal conductivity on the whole will be increased. With this in mind, for spacecraft structures with insulated coatings the basalt superthin fibre-canvas BSTF "MINOL" [11] with the fibres diameter d_f = 0.5 to 3.0 μm and the fibres length in the canvas l_f =50-60 mm is selected.

To determine the thickness of the insulation coating for the spacecraft, the average coefficient of the TPM thermal conductivity, is required and this depends on the porosity of the fibrous structure. Neglecting the Stefan-Boltzmann radiant heat loss, and heat loss for calm air in the insulating material pores, its thermal conductivity coefficient can be written based on the mixtures principle

$$\lambda(n,T) \cong \lambda_{TPM}^{ev} = n\lambda_{air}^{ev} + (1-n)\lambda_{BF}^{ev}, \tag{8}$$

where the average values of the coefficients λ_{BF}^{ev} = 0.0587 W/(m·K) and λ_{air}^{ev} = 0.0413 W/(m·K) for temperature range from 60°C to 400°C are defined in [3]. Then, for material with porosity n = 0.92 to 0.94, the average coefficient of TPM thermal conductivity is λ_{TPM}^{ev} = 0,0425 W/(m·K).

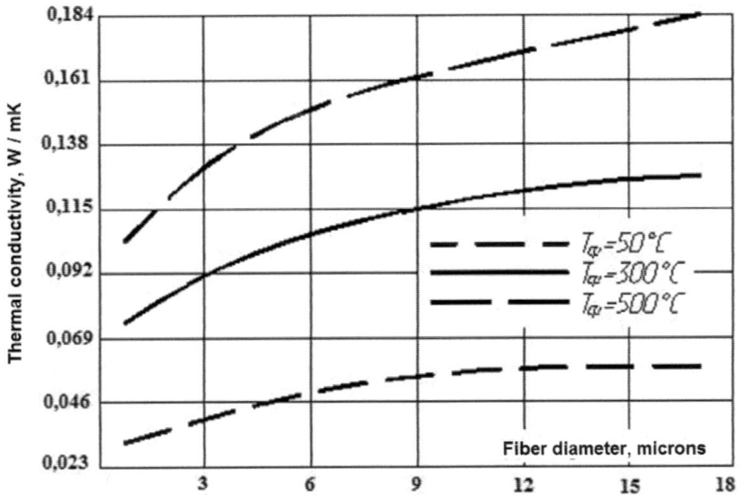

Figure 5. The dependence of the basalt fibre thermal conductivity of their diameter at different temperatures

Considering that most of rockets and spacecraft have a cylindrical form, and accepting that the heat flow along the tube length of q_1=150 W/m is inconsequential, the required HIC thickness of the basalt fibres can be pre-calculated using the formula

$$\frac{q_1}{\pi d} = \frac{2}{d}\frac{\lambda_{TPM}(T_1 - T_2)}{ln\frac{d+2h_{TPM}}{d}}, \tag{9}$$

where d is the rocket or spacecraft outer diameter, d = 60 mm; T_1=400°C – the temperature on the inner wall of rockets and spacecraft; T_1=60°C – the temperature on the outside of the basalt cover; and h_{TPM} is the thickness of the heat insulation of the spacecraft. By substituting λ_{TMP}^{ev} from (8) into formula (9), we will define the thickness of spacecraft structure heat insulation: h_{TPM} = 25mm.

To increase the compressive strength of the high-porosity fibrous TPM, the high-temperature mineral binder is used, which fixes the fibres at the sites of their contacts. The selected binder, aluminium oxide $A1_2O_3$, is derived from salts of aluminium sulphate in aqueous solution of ammonia, presented in hydro mass in a colloidal condition, which increases the liquid pulp and the filtrate viscosity.

The hydro mass prescription composition, for the manufacture of thermal insulation products, is as follows: 0.95 kg of the short basalt fibre, 0.4 l of 27 % $Al_2(SO_4)_3$ $18H_2O$ solution of aluminium sulphate, and 0.07 l of 25 % $6NH_4OH$ aqueous ammonia solution. The formation of aluminium oxide occurs in the process of drying and heat treatment of the product at a temperature of 300°C. In the process of filtration deposition the amount of pulp acidity must be strictly controlled: it is necessary to perform hydrolysis and deposition of the fibres at pH = 7.6 to 8.4. In this case, one can ensure the calculated amount of aluminium oxide in the heat insulation material in the form of binder, while maintaining a sufficiently high filtration rate.

During fibre-free filtration deposition from pulp, the time for the accumulation of a sediment layer of height up to 100 mm (to the cylindrical bushing) is not less than 90 min, according to formula (4), and in this case, the problem of material thickness equidensity and the product surface geometry is not fully solved. The dynamics of the hydro mass column reduction as a function of time is presented in figure 6.

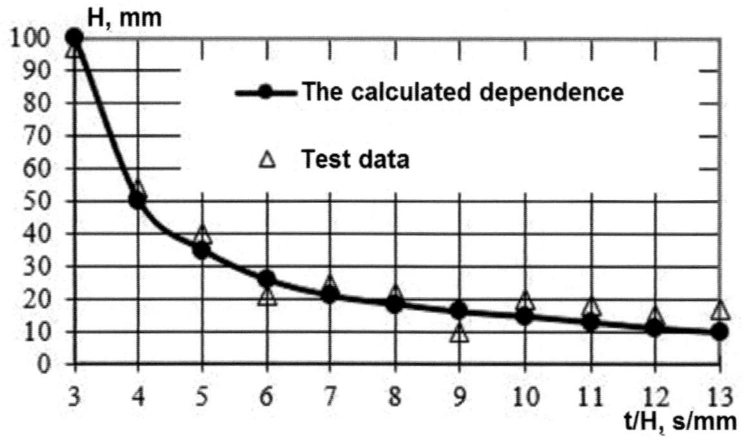

Figure 6. The dependence of hydro mass column height of the time during free filtering

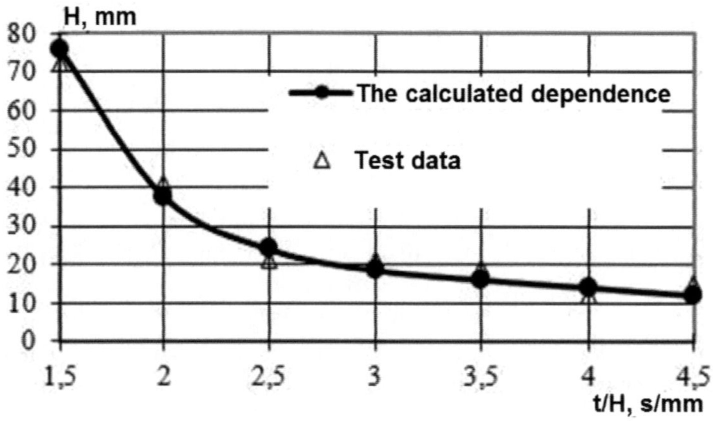

Figure 7. The dependence of hydro mass column of the time with vacuum treatment and pre-moulding

The efficiency of the process increases significantly as a result of filtrate chamber vacuum treatment (figure 7). In this case, the value of Δp in formula (3) is a free parameter. The pressure of vacuum treatment should be chosen in such a way that the thickness and density of the sediment layer would not be less than the designed (accepted) values. In this case, the fibre filtration deposition time should be less than that of the hydro mass layering on liquid and solid phases. In fact, equation (3) can be considered as the time control law for sediment layer filtration and density for the manufacture of products from fibrous heat insulating material with a predetermined thickness h_{layer}.

Cylindrical rings, manufactured by fibre-free filtration deposition from pulp, with wall height of more than 50-60 mm, have a material density variation in respect of height of up to 12-15%. To reduce the material inhomogeneity, it is necessary to apply a method of liquid filtration with the filtrate chamber vacuum treatment, and then finished pre-moulding of the sediment layer by means of punch, to form the final geometric size and shape of the product. Constant pressure difference across the sediment layer under vacuum treatment can be steadily maintained in the range of Δp=50 to 70 kPa.

To reduce the manufacturing time of the long cylindrical shells made of the short basalt fibres by liquid filtration, it is necessary to apply a moulding process using the plane generating cylinder shown in figure 8. Initially, for a short time (7-10 min), the deposition of the main volume of fibres takes place, and then, without interrupting the process, the mandrel–punch is introduced, and fibre filtering and the TPM lock moulding ends. With such a scheme of filtering, the pulp column height and the product manufacturing time decreases sharply. Additionally, the design of a manufacturing jig is simplified significantly.

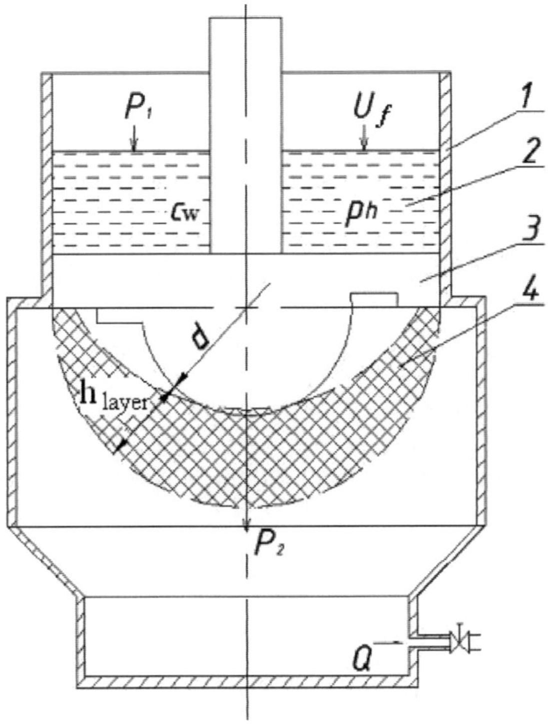

Figure 8. The scheme of cylindrical shells moulding made of the short basalt fibres:
1 – camera; 2 – pulp; 3 – punch; 4 – the initial layer of fibre sediment

To determine the actual density of the "shell" product obtained by the filtration deposition method, measurements of mass and geometric parameters of the product were carried out. Products from four different batches were taken. The geometrical parameters were almost equal, with minor errors, resulting in figure 9: D_{TPM}=110±0.2 mm; d=60±0.2 mm; L_{TPM}=255±0.1 mm; h_{TPM}=25±0.1 mm. The average weight of the shells was found equal to 0.1327 kg, the material density ρ_{TPM}=156±3.4 kg/m^3, and the fibrous composition porosity was 94 %.

Figure 9. View of the assembled heat insulation basalt shells for rockets and spacecraft

4. Conclusions

The proposed mathematical model for the free filtration deposition of short fibres from liquid pulp describes the dynamics of the heat insulating product formation, and determines rational parameters of the technological process of their manufacture.

To improve product quality and increase the productivity of the deposition process for short basalt fibre filtration, it is necessary to use the filtrate chamber vacuum treatment and sediment layer finish pre-moulding to obtain the final geometric size and shape of the product.

To increase the compressive strength of the highly porous fibrous rings and cylindrical shells, it is necessary to use a mineral binder of 5-7 % by weight, based on oxide $A1_20_3$, that fixes the basalt fibres in the sites of their contact.

The cylindrical shells obtained by the filtration deposition method from the short basalt fibres meet the required quality.

References

[1] Лозино-Лозинского ГЕ, Братухина АГ 1997 *Авиационно-космические системы: сб. ст.* Изд-во МАИ [Lozino-Lozinskii GE, Bratukhin AG eds 1997*Aerospace systems: collected articles* MAI Publ, Moscow (in Russian)]

[2] Тарасов ВА, Смирнов ЮВ, Тимофеев МП, Филимонов АС 2007 Режимы фильтрационного осаждения элементов теплозащиты РКТ, *Полет. Общероссийский научно-технический журнал* **5** 52–55 [Tarasov VA, Smirnov YuV, Timofeev MP and Filimonov AS 2007 Modes of Filtrational Sedimentation of Heat-Insulation Elements of Space-Rocket Technics *Polet Flight All-Russian Scientific and Technical Journal* **5** 52–55 (in Russian)]

[3] Комков МА, Баданина ЮВ, Тимофеев МП 2014 Разработка и исследование термостойких покрытий трубопроводов из коротких базальтовых волокон *Инженерный журнал: наука и инновации* **2** [Komkov MA, Badanina YuV and Timofeev MP Developing and testing heat-resistant pipelines coatings made of short basalt fibers 2014 *Engineering Journal: Science and Innovation* **2** (in Russian)] http://engjournal.ru/catalog/machin/hidden/1203.html

[4] Филимонов АС, Тарасов ВА, Комков МА, Моисеев ВА, Тимофеев МП, Герасимов НВ 2012 Экспериментальный анализ свойств перспективных

теплоизоляционных материалов машиностроения, полученных методом фильтрационного осаждения *Инженерный журнал: наука и инновации* **9** [Filimonov AS, Tarasov VA, Komkov MA, Moiseev VA, Timofeev MP and Gerasimov NV 2012 Experimental analysis of the properties of advanced thermal insulation materials engineering, obtained by filtration deposition *Engineering Journal: Science and Innovation* **9** (in Russian)] http://engjournal.ru/catalog/machin/rocket/382.html]

[5] Филимонов АС, Тарасов ВА, Комков МА, Моисеев ВА, Тимофеев МП, Герасимов НВ 2012 Влияние связующих на свойства новых теплоизоляционных покрытий с использованием стеклянных микросфер осаждения *Инженерный журнал: наука и инновации* **9** [Filimonov AS, Tarasov VA, Komkov MA, Moiseev VA, Timofeev MP and Gerasimov NV 2012 Effect of binders on the properties of the new thermal insulation coatings using glass microspheres *Engineering Journal: Science and Innovation* **9** (in Russian)] http://engjournal.ru/catalog/machin/rocket/383.html]

[6] Тимофеев МП 2007 *Разработка и исследование фильтрационной технологии изготовления изделий из волокнистых неорганических материалов: автореф* дис. канд. техн. наук МГТУ им. НЭ Баумана М [Timofeev MP 2007 *Development and research of filtration technology the manufacture of fibrous inorganic materials* diss. Bauman MSTU, Moscow (in Russian)]

[7] Леонтьев НЕ 2009 *Основы теории фильтрации* М: Изд-во ЦПИ при механико-математическом факультете МГУ [Leont'ev NE 2009 *Fundamentals of the theory of filtration* Publishing House of the CPI at the Mechanics and Mathematics Faculty of Moscow State University, Moscow (in Russian)]

[8] Collins RE 1961 *Flow of Fluids Through Porous Materials* New York, Reinhold Publishing Corp

[9] Muskat M 1946 *The Flow of Homogeneous Fluids Through Porous Media* Ann Arbor, Michigan, JW Edwards Inc

[10] Джигирис ДД, Волынский АК, Козловский ПП и др. 1980 Основы технологии получения базальтовых волокон и их свойства *Базальтовые волокнистые композиционные материалы и конструкции: сб. науч. тр.*: Наукова думка, Киев 54-81 [Dzhigiris DD, Volynskii AK, Kozlovskii PP *et al* 1980 Fundamentals of technology of basalt fibers and their properties *Basalt fibre composite materials and structures: collected papers* Naukova dumka Publ, Kiev 54-81 (in Russian)]

[11] *Базальтовое супертонкое волокно «MINOL»* Завод БСТВ «MINOL»: сайт компании [*Super thin basalt fiber "MINOL"* The plant BSTV "MINOL" company website] http://uteplitel-minol.ru/holst/

Influence of the surface cleaning methods of composite products on the quality of coverings

AN Korolev, AN Alexeev and VA Tarasov

BMSTU, 5, 2nd Baumanskaya St., Moscow, 105005, Russia.

dekanatsm2010@mail.ru

Abstract. Studies have been conducted relating to a priority area – the sustainable use of natural resources, based on the suggested design arrangement of the operating module, with counter-current motion of washing liquid in the system of a distributed wash. The assignment of permittable concentrations of chemical compounds to be washed-off on elements as well in wash tanks is suggested. Regularities as to the change of electrolyte concentration on element surface and in distributed wash system are established. Results are presented of numeric calculations of the spray rinse process of the elements having blind and threaded holes filled with liquid impurities. Spray rinse of elements with blind holes, and blind threaded holes are compared. Substantiation of jet velocity and recommendations as to the minimal diameter of the washing of thread holes which ensures efficient removal of scum, is carried out. The studies make exposure of components at spray rinse more specific. The technique for assignment of cross-feed of parts towards multi-nozzle heads which ensures effective scum removal was substantiated.

1. Introduction

The coating application for composite materials requires on application of in-process item cleaning by rinsing, accounting for up to 10% of labour for item fabrication. Until now, processes of part surface cleaning remain energy and resource consuming, requiring a comparatively large amount of time and production area resources. Water-consumption during the cleaning process may reach 4 m^3 per 1 m^2, whereas up to 1 m^3 of liquid waste per 1 m^2 is generated (which is characteristic for the electroplating industry). On a national scale, this results in the generation of a substantial volume of waste water (up to 10 million m^3 annually) with a small concentration of ions and salts of heavy metals (0.01 to 0.1 gr/l). Due to operations at centralised water treatment facilities being inefficient, thousands of tons of non-ferrous metals are lost irrecoverably [1, 2].

With this in mind, the fact that many companies use component rinsing, this problem should be treated as urgent [1].

According to Henley [2], the common fault is the incomplete and ineffective wash between two adjacent process stages, which results in excessive build-up of electrolyte scum and deterioration of processing quality, namely when washing complicated parts. These parts contain a number of projections, undercuts, narrow slots as well as deep (up to 10-12 mm of depth) openings of small diameter (1-2 mm), including threaded holes.

Currently, recommended wash modes are restricted either by a minimal wash period at spray rinsing or by a number of replacements (per hour) of wash solution in a container. Such an approach

IOP Conf. Series: Materials Science and Engineering **153** (2016) 012010 doi:10.1088/1757-899X/153/1/012010

does not permit a reasonable consumption of water, nor for the extraction of important chemical elements from the wash solutions and from the washed layer [3].

Spray rinse has a number of advantages as compared to other techniques for item cleaning: these are efficiency and economy [4, 5]. Creating a technique for the selection of modes for the spray-dynamic wash of complicated parts is rather a sophisticated task which has not been resolved until now [6].

The purpose here is to minimise loss of water and valuable chemical compounds, as well as to minimise the energy consumption of the water treatment facilities.

2. Statement of the problem

As the basis for securing the recovery of water and valuable chemical compounds, we offer a design arrangement of an operating module of electroplating facilities (figure 1) with a counter-current flow of wash fluid in a distributed wash system. The idea of such a solution is set out in [1].

The operating module contains the required amount of process and rinse tanks which ensure the operation of primary processing and cleaning of the part's surface from residues of water solution, metal salts, acids and alkali used.

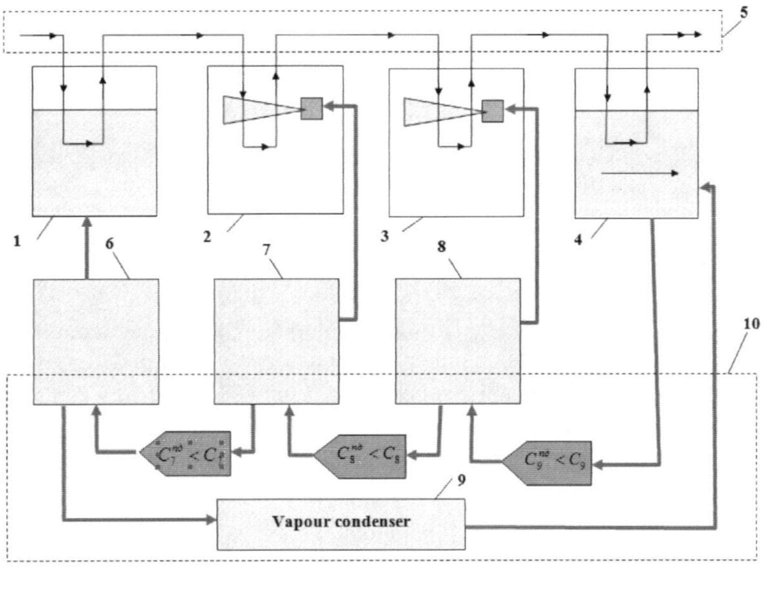

Figure 1. Diagram showing the operating principle of electroplating manufacture: 1 – process tank, 2, 3 – spray rinse tanks with jet formation elements, 4 – flowing water dip rinse tank, 5 – path of part direct-current flow, 6 – concentrator-evaporator tank for servicing process, 7, 8 – tanks for servicing wash, 9 – vapour condenser, 10 – path of counter-current flow of wash solution

In order to solve the task the operating module is divided into two levels – upper and lower. In the upper level the path of part direct-current flow 5 is realised, where parts being processed are sequentially delivered to work tanks 1-4 of each stage. In process tank 1, which is filled with electrolyte or with chemical compound solution, electrochemical processing of parts takes place (degreasing, pickling treatment, application or forming of the relative coating, passivation, dyeing of surfaces, chrome-plating, etc.), in tanks 2 and 3 they are spray-rinsed and in tank 4, the parts are finally dip-washed.

Vessels and work tank servicing units are placed in the lower level. Vessels 7-9 with washing solution service rinse tanks 2-4. For each rinse stage the maximum permissible concentration level, K_i^{Σ}, of electrolyte is used in the wash solution, where i is the stage number. As the washed area of parts increases, the concentration, K_i, of electrolyte in the wash solution grows, while remaining sufficient for ensuring quality of rinsing.

At the moment when the maximum permissible concentration, K_i^{Σ}, is reached, the pumping of wash solution from vessel i to the vessel of the preceding rinse stage, $i-1$, takes place. From vessel 7, the wash solution containing scum after rinse 1 is transferred to the concentrator-evaporator 6, where the concentration of electolyte necessary for the operation of process tank 1 is reached.

Vapour to be obtained using the concentrator-evaporator 6 is supplied to the vapour condensing unit 9 through a filter system. As a result, water is obtained, which is returned to the tank 4. Thus, the chemical compounds and water from the wash solution are not lost irrecoverably at water treatment facilities, but are returned to the operation thereby reducing consumption of water and compounds.

3. Simulation of inter-relation of states of washed surface of parts and wash solution in operating module vessels

In [7], correlations are obtained for the assignment of maximum permissible concentration of wash solution in vessels, which is critical to the control of the operating module.

Thus, the concentration of the component being rinsed on the part surface after tanks with N stages of rinsing can be represented with the single recurrent formula:

$$C_{n+1}^{\Sigma} = \left[\sum_{n=1}^{N} \left(K_{n-1}^{\Sigma} - K_{n}^{\Sigma} \right) \gamma^{N-n+1} \right] \beta + K_{n+1}^{\Sigma}, \qquad (1)$$

where $\beta = \exp(-a_{spr} t_{spr})$, $\gamma = \exp\left[-a_{dip}\left(t_{gt} - N t_{spr}\right)\right] \approx \exp\left[-a_{dip} t_{gt}\right]$, a_{spr} and a_{dip} are the speed of reduction in concentration of solution being rinsed on component surface at spray-dynamic rinse and at dipping, respectively, where N is the number of spray rinse stages and n is the number of the particular rinse stage. K_n^{Σ} is the maximum permissible concentration of rinsing component in the wash fluid, and C_{N+1}^{Σ} is the resulting concentration of that component on the surface of the part after being washed.

The design task for the electroplating processing operating module is to define all the values for C_n^{Σ} and K_n^{Σ}.

To find the terms in equation (1) which are not explicitly defined, the following balance equations can be defined. These balance the change in the weight of the chemical compound in the vessel, and the critical concentration of compound in the rinsing vessels of wash fluid of the rinse stages which correspond to vessels of previous stages.

$$\begin{aligned} K_1^{\Sigma} &= \left(C_0^{\Sigma} - K_2^{\Sigma}\right)\alpha + K_2^{\Sigma} \\ K_2^{\Sigma} &= \left(C_1^{\Sigma} - K_3^{\Sigma}\right)\alpha + K_3^{\Sigma}, \\ K_3^{\Sigma} &= \left(C_2^{\Sigma} - K_{\text{тв}}\right)\alpha + K_{\text{тв}} \end{aligned} \qquad (2)$$

where $\alpha = 1 - \exp[-\bar{G}t\delta/v]$; K_{TB} is the concentration of rinsed component in process water that may be accepted as equal to zero; v is the volume of washing fluid in vessel; $\bar{G}$ is the area washed in a time unit t; and δ is the thickness of the layer containing wash solution.

The correlations presented allow the calculation of the critical concentrations of the chemical wash component on parts and in wash vessels, without which it is impossible to plan the operation of the module. In [8] the suggested method is experimentally tested via the example of the nickel plating operation.

4. Numerical simulation of the spray-dynamic rinse of openings with various configurations

The rinse of flat components is resolved analytically in work [6]. For the rinsing of parts with more complicated shapes the analytical approach becomes challenging because of the factors involved. Numerical calculations [9, 10] were performed using ANSYS [11], to obtain the information necessary to develop techniques for the assignment of operation modes and conditions for the spray-dynamic rinse of shaped components.

The modelled axially symmetric area included the following objects (figure 2):

- A water jet with diameter d_c, continuously flowing with speed v_c,
- Fluid containing scum which fills into the closed threaded opening, with water column length l_0 and diameter d_0, (the density, viscosity, and the tensile strength of the water column and the fluid containing scum are assumed to be identical);
- An area above front surface of part, not initially filled with medium, and assigned for studying the fluid discharge from the closed hole.

The tensile strength of the water was taken as 28 MPa, according to data for distilled water. Calculations are made within range v_c = 3 to 7 m/s for metric thread with diameter 3, 4, and 5 mm.

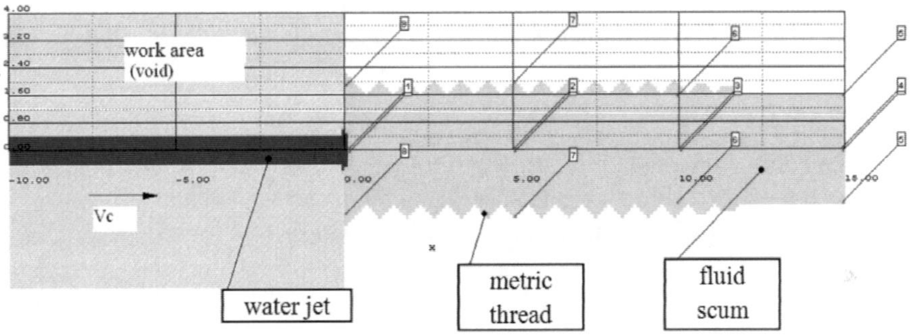

Figure 2. Model geometry

5. Discussion of the results for the various configurations.

5.1. *Investigation of spray rinse for openings of various configurations*

In order to study the physical properties and to justify the selection of particular modes of spray-dynamic rinse of components with closed threaded holes, a comparative analysis was performed of the cleaning of closed holes with and without a thread (figure 3). As an example, let us consider openings with diameter 4 mm (D4) and threaded hole M4, subjected to a water jet from multi-nozzle heads with velocities v_c = 3, 5 and 7 m/s.

As a result of numerical modelling of the axially symmetric process of liquid jet penetration which jet continuously penetrates to closed openings filled with liquid impurities the principle regularities were established (figure 4).

The best conditions for the rinsing of blind openings without threads are ensured by a quasi-stationary mode of jet penetration into the scum volume, which is realised under the following conditions: $d_o > d_c$ (d_o — diameter of opening; d_c — diameter of jet) and velocity of sprinkling of water jet from multi-nozzle heads, v_c, is greater than 5 to 7 m/s. The linear dependence of the relative volume of residual impurities on the volume of residual impurities is defined as follows

$$\mu(t) = \frac{V_{\text{res}}}{V_r} \cdot 100\ \%$$

where μ is the relative volume of residual impurities, V_{res} is the volume of residual impurities at an arbitrary point of time t, and V_r is the volume of residual impurities at t = 0.

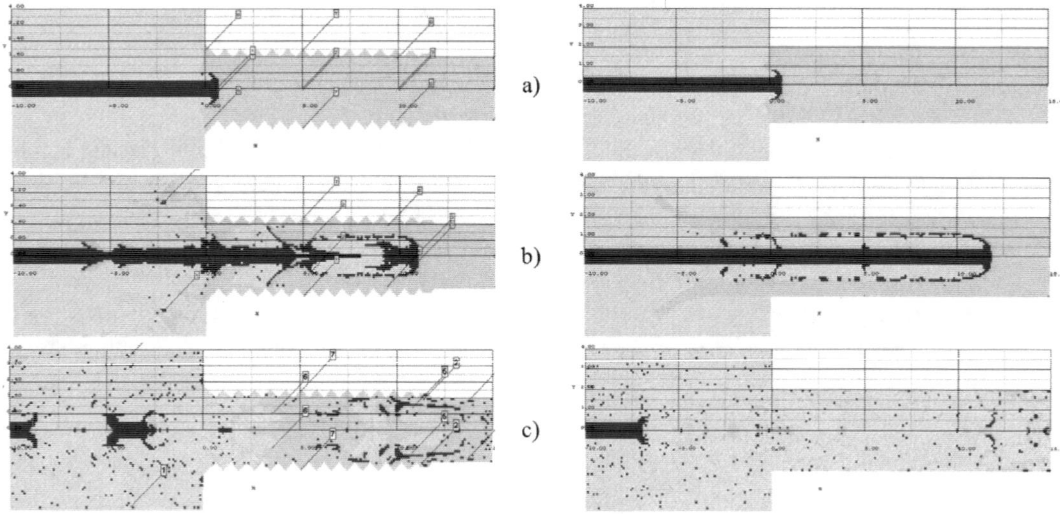

Figure 3. The volume of residual impurities following hydro-treatment of a hole with water jet velocity v_c = 3 m/s, at different moments of time:
(left) with metric thread, (right) without metric thread,
a) t = 0.5 ms; b) t = 10 ms; c) t = 30 ms

With an increase in discharge rate, the purification rate increases. For a quasi-stationary mode, around 90 % of impurities are removed after a period of around 25 ms. When the jet reaches the bottom, the function linearity $\mu(t)$ is violated, resulting from side impacts of impurity being washed-off along the jet moving down the hole. At this point the rate of impurity removal slumps. After around 40 ms the hole can be considered purified.

As a result of the complex configuration of threaded hole, the quasi-stationary mode of jet penetration into impurity is reached within a short period of time – around 5 ms. Further hydro-treatment is accompanied with uninterrupted side impacts of impurity being washed off whilst the jet moves inside the hole, with a vortex of flows and the slow washing of hard-to-reach thread grooves. The consequence of the side impacts described above is the need for a considerably increased cleaning time (up to 90 ms and more), and performance is reduced by a factor of two or three.

5.2. *Influence of thread hole diameter on the hydro-treatment mode*

Figure 5 demonstrates the characteristics of hydro-treatment modes for holes M3, M4, M5 at flow velocity v_c = 5 m/s.

The cases for water flow velocity $v_c = 3, 5$ and 7 m/s, and for holes M3, M4 and M5 are presented in figure 6. Because of the complicated surface configuration for holes with diameter, d_o less than 3 mm, it is difficult or impossible to use this method to model the cleaning. For larger diameter holes we can see a considerable increase in performance of the hydro-treatment process, irrespective of water velocity ($d_o = 0.8$ mm). Use of the spray rinse method when the "index of method applicability" coefficient, $K_e = d_o/d_c$ exceeds 5 can be considered practical.

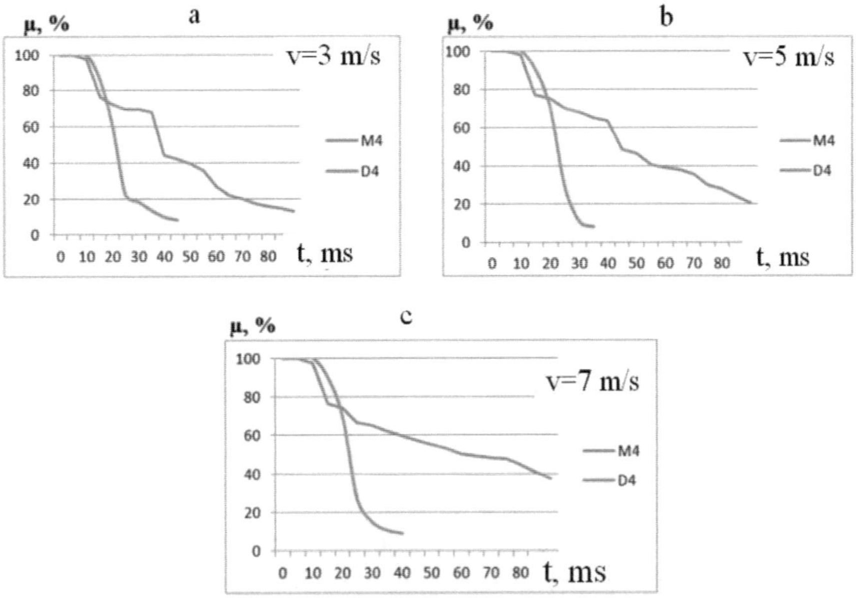

Figure 4. Volume of residual impurities as a function of time, for different water discharge velocities: a) 3 m/s; b) 5 m/s; c) 7 m/s

5.3. *Influence of water flow velocity on the hydro-treatment mode*
Figure 7 demonstrates characteristics of hydro-treatment modes for hole M5 at different flow velocity.

For the purposes of increasing performance of spray rinse with an increase of threaded hole diameter, the importance of the correct choice of water supply velocity is more significant (figure 8). It can be seen that the smallest values of process performance were obtained at a water jet velocity of 5 m/s, and the largest were at a reduced velocity of 3 m/s. Thus, the conclusion is that the relationship between process performance and water jet velocity is not linear and additional experiments are required. One can assume that with such complicated hole configurations, the performance values will be unpredictable.

6. Conclusions

A design arrangement of operating module for electroplating industry which ensures minimal consumption of water, chemical compounds as well energy is proposed. A method of assigning critical concentrations of component to be washed on the parts and in wash vessels was suggested.

The mathematical simulation of component spray-dynamic wash process for components with threaded holes allowed: the development of new recommendations as to the assignment of wash modes for threaded holes; the performance of a comparative evaluation of hydro-treatment velocity between threaded holes and closed holes without thread; an assessment of the cleaning method for small-diameter threaded holes.

The work demonstrated an increase in economy and performance of spray-dynamic wash method, for a reasonable consumption of water resources and valuable chemical elements when cleaning items,

IOP Conf. Series: Materials Science and Engineering **153** (2016) 012010 doi:10.1088/1757-899X/153/1/012010

and a reduction in the volume of waste water. Each of these factors demonstrates increased environmentally friendly manufacturing.

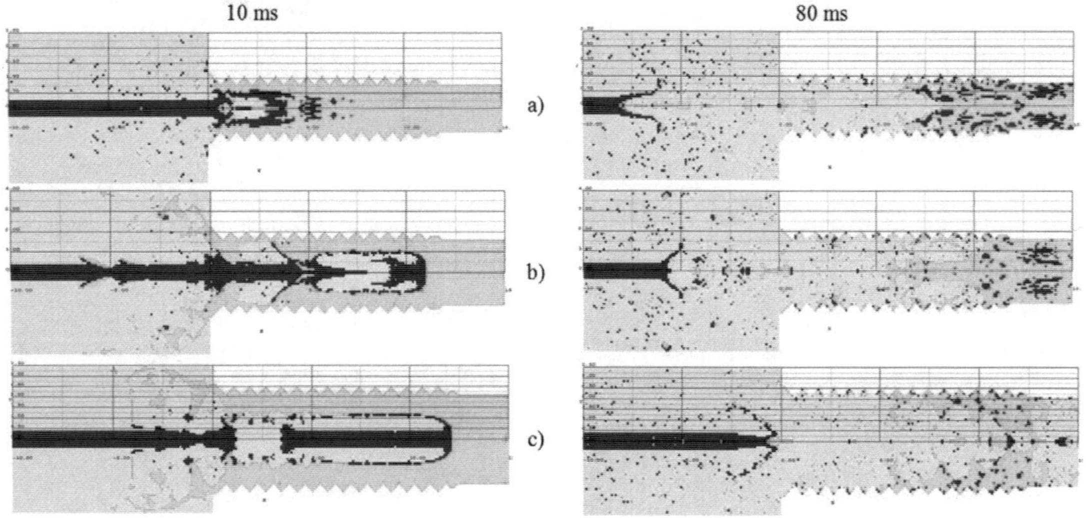

Figure 5. The volume of residual impurities following hydro-treatment with water jet velocity v_c = 5 m/s, at different moments of time: (left) t = 10 ms, (right) t = 80 ms, of threaded holes a) M3; b) M4; c) M5

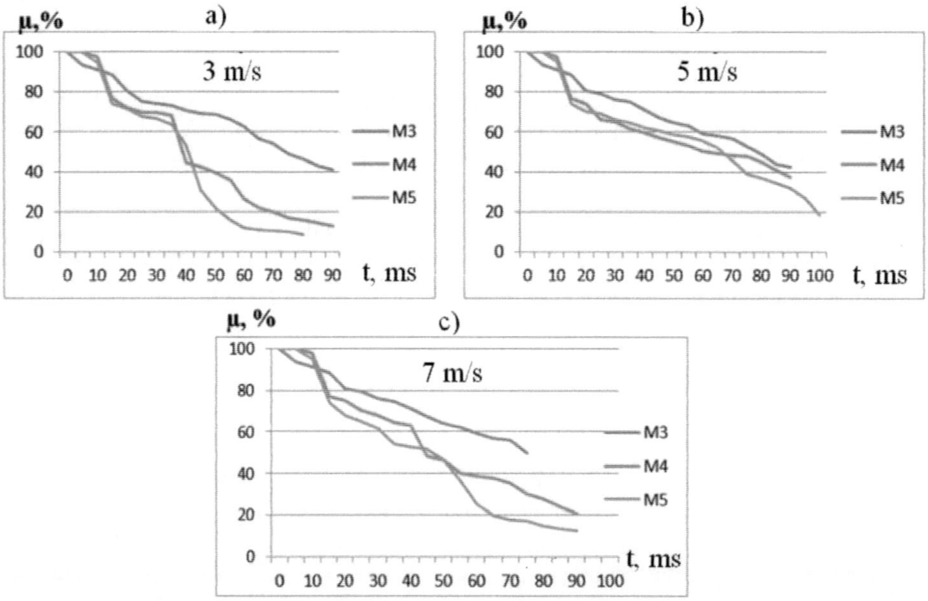

Figure 6. Volume of residual impurities as a function of time, in openings M3, M4 and M5, for water discharge velocities: a) 3 m/s; b) 5 m/s; c) 7 m/s

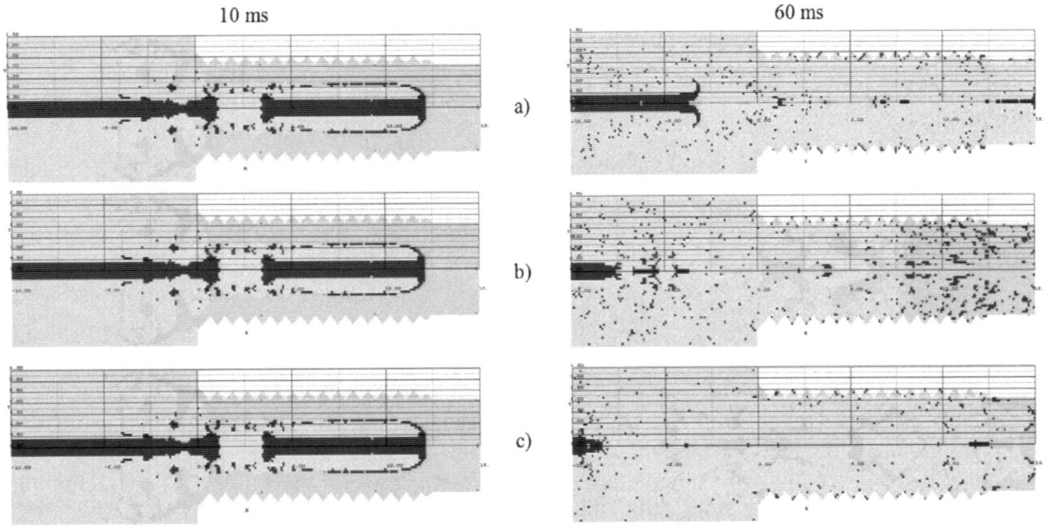

Figure 7. The volume of residual impurities following hydro-treatment of hole M5 at different moments of time: (left) t = 20 ms, (right) t = 60 ms, with water jet velocity, a) v_c = 3 m/s; b) 5 m/s; c) 7 m/s

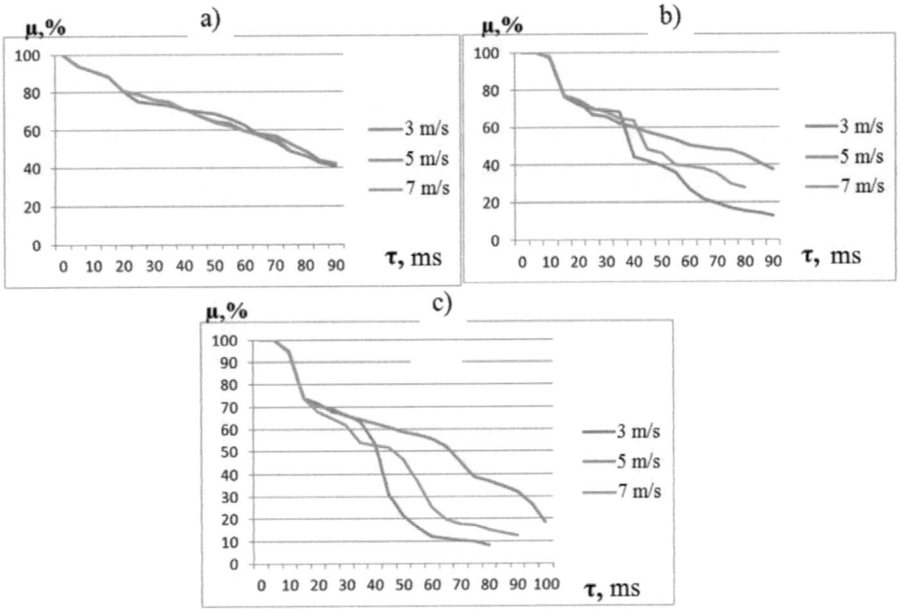

Figure 8. Volume of residual impurities as a function of time, for different water discharge velocity, at the diameter of holes: a) M3; b) M4; c) M5

References

[1] Alekseyev AN 1997 Концепция развития и пути создания современного гальванического оборудования *Новые промышленные технологии* **3** 2–12 [Concept of development and ways to make advanced electroplating equipment *New Industrial Technologies* **3** 2–12 (in Russian)]

[2] Henley V 1986 *Anodic oxidation of aluminum and its alloys* Pergamon Press, Oxford

[3] Vinogradov SS 2007 *Промывные операции в гальваническом производстве* Globus, Moscow [*Washing operations in electroplating facility* Moscow Globus (in Russian)]

[4] Alekseyev AN and Tarasov VA 2003 Разработка и внедрение струйной пакетной технологии для очистки поверхностей деталей на предприятиях машиностроительного комплекса *Инженерный журнал* **12** 6–10 [Development and introduction of spray package technology to clean part surfaces in enterprises of mechanical engineering sector *Engineering magazine* **12** 6–10 (in Russian)]

[5] Tarasov VA and Galinovskiy AL 2013 Проблемы и перспективы развития гидроструйных технологий ракетно-космического машиностроения *Инженерный журнал: наука и инновации* **3** 23–27 [Problems and perspectives of development of water-jet technologies in rocket-and-space engineering *Engineering Journal: Science and Innovation* **3** 23–27 (in Russian)]

[6] Tarasov VA Alekseyev AN, Boyarskaya RV and Korolev AN 2010 Повышение эффективности гальванохимической обработки деталей ЛА *Общероссийский научно-технический журнал «Полет»* **3** 42–46 [Increase in performance of galvanic-chemical processing of components aircraft *All-Russian scientific-technical magazine "Polet"* **3** 42–46 (in Russian)]

[7] Tarasov VA, Korolev AN and Boyarskaya RV 2011 Бессточная многоступенчатая промывка элементов машин от химических загрязнений *Вопросы оборонной техники* **16**(1,2) 47–51 [Closed-cycle multi-stage wash of machine elements from chemical contaminations *Questions defense equipment]* **16**(1,2) 47–51 (in Russian)]

[8] Korolev AN, Tarasov VA and Alekseyev AN 2012 Экспериментальный анализ изменения концентрации раствора в операционных модулях травления и обезжиривания с многоступенчатой промывкой деталей *Инженерный журнал: наука и инновации* **9** 30–34 [Experimental analysis of variation of solution concentration in operating modules of pickling treatment and de-greasing with multi-stage wash of parts *Engineering Journal: Science and Innovation* **9** 30–34 (in Russian)]

[9] Kazakova OI and Kolpakov VI 2012 Численное моделирование гидроабразивной резки листовых заготовок из алюминиевых сплавов *Известия высших учебных заведений. Машиностроение* **7** 56–60 [Numeric simulation of water-jet cutting of sheet stock of aluminium alloys *Proc Higher Educational Institutions. Machine building* **7** 56–60 (in Russian)]

[10] Abashin MI, Barzov AA, Galinovskiy AL, Kazakova OI, Kovalev AA, Kolpakov VI, Mulyar SG, Novozhilov SA and Sysoyev NN 2011 Численное моделирование гидрофизических процессов в зоне ударно-динамического взаимодействия ультраструи жидкости с твердотельной мишенью *Изд-во: МГУ им. М.В. Ломоносова* **4** 35–40 [Numeric simulation of hydrophysical processes in the area of impact-dynamic interaction of fluid ultra-jet with the rigid-body target *Lomonosov Moscow State University MGU* **4** 35–40 (in Russian)]

[11] Tarasov VA, Kolpakov VI, Korolev AN and Baskakov RV 2011 Численное моделирование процесса струйно-динамической промывки деталей с глухими отверстиями *Вестник МГТУ им. Н.Э. Баумана* **4** 34–41 [Numeric simulation of spray-dynamic wash process for components with closed holes *Vestnik BMSTU* **4** 34–41 (in Russian)]

IOP Conf. Series: Materials Science and Engineering **153** (2016) 012011 doi:10.1088/1757-899X/153/1/012011

Lifelong modelling of properties for materials with technological memory

AP Falaleev[1], VV Meshkov[2], AA Vetrogon[2], SV Ogrizkov[2] and AV Shymchenko[2]

[1] Peter the Great St. Petersburg Polytechnic University, Russian Federation;
[2] Sevastopol State University, Russian Federation.

a_falaleev@mail.ru

Abstract. An investigation of real automobile parts produced from dual phase steel during standard periods of life cycle is presented, which considers such processes as stamping, exploitation, automobile accident, and further repair. The development of the phenomenological model of the mechanical properties of such parts was based on the two surface plastic theory of Chaboche. As a consequence of the composite structure of dual phase steel, it was shown that local mechanical properties of parts produced from this material change significantly their during their life cycle, depending on accumulated plastic deformations and thermal treatments. Such mechanical property changes have a considerable impact on the accuracy of the computer modelling of automobile behaviour. The most significant errors of modelling were obtained at the critical operating conditions, such as crashes and accidents. The model developed takes into account the kinematics (Bauschinger effect), isotropic hardening, non-linear elastic steel behaviour and changes caused by the thermal treatment. Using finite element analysis, the model allows the evaluation of the passive safety of a repaired car body, and enables increased restoration accuracy following an accident. The model was confirmed experimentally for parts produced from dual phase steel DP780.

1. Introduction

Numerical methods for engineering calculations, such as Finite Element Analyis, have become the most popular tool for engineering design; however, these methods can only give an adequate simulation of real processes where the material behaviour is adequately modelled. That is why it is so important to have an appropriate description of the properties of new materials, and it is for this reason that the implementation of new materials for designs using old calculation methods of or old models can lead to unpredictable results. Materials with technological memory are the most difficult type in respect of the implementation of their properties within a calculation, because the material characteristics change considerably at each step of the technological operation. Usually these changes of properties become apparent locally or are direction of loading dependent, so after a technological treatment the properties such of materials are anisotropic and heterogeneous. Complex phase steels, widely used as the basic metal for automotive vehicle body manufacture, are representative examples of such materials.

IOP Conf. Series: Materials Science and Engineering **153** (2016) 012011 doi:10.1088/1757-899X/153/1/012011

Dual phase steel has a composite microstructure that consists of grains of martensitic phase and perlite phase, and demonstrates significant work hardening under plastic deformation. This structure is formed by quenching of steel at the temperature range of the critical temperatures of austenite transformation. The main drawback of this microstructure is that it can be inadvertently altered during the repair, as a result of heating, leading to reduced performance properties. Such heating or changing of the operational temperature during restoration, as is typical of various repair methods, has a significant influence on the microstructure of different kinds of high strength steels [1-5]. The great benefit of such metals is the capacity to dissipate large amounts of energy through plastic deformation during impact, which is a consequence of the high level of hardening. At any given point in the life of a dual phase steel component, the mechanical properties depend on all plastic deformations accumulated during previous operations.

A phenomenological approach is the most common tool for simulating the properties of dual-phase steels. The Chaboche kinematic hardening model is widely used to describe the behaviour of steel [4]. This approach combines isotropic hardening and non-linear models that take into account the Bauschinger effect [5, 6]. The calculation of mechanical behaviour after technological operations is a necessary and most difficult task. The degradation of dual-phase steel properties during repair procedures requiring heating are studied in papers [7, 8]. Research into the alteration of dual-phase steel hardness in the heat affected zone during welding is presented in [9], where the reduction of strength following cooling is attributed to the partial release or martensite; however the majority of research interest is in the influence of thermal effects on the dual phase steel during new material stamping. During the life of a real engineering component, it accumulates a lot of technological operation influences involving plastic deformation, repair, welding, heating, or straightening. One such example is a side member of car body, initially manufactured by stamping and welding, subsequently plastically deformed during an accident, and then repaired by pre-heating then stretching the pre-heated dent. The problem is to predict the real mechanical properties of this part, and how to simulate its behaviour under subsequent plastic deformation.

2. The aim and strategy of the investigation

The aim of this article is to develop algorithms, models and equations for calculating the mechanical properties of materials with technological memory following technological treatments, as in the example of the side member of a car body made from dual phase steel, following the consecutive processes of production, plastic deformation, repair and further new plastic deformation.

The most critical technological factors causing microstructural changes, and therefore changes in mechanical properties as it is composite structure with nonstable microstructure, are: accumulated plastic deformation, taking account of the history of the deformation directions; deformation rate; current temperature of the metal; history of thermal treatment and relaxation period.

The strategy of this work is to develop:

1. A phenomenological model for the mechanical properties of dual phase steel, based on Chaboche two surface plasticity, by adding
2. Empirical relations for the stretch behaviour of pre-compressed heated metal,
3. Empirical relations for the evolution of mechanical properties of dual phase steel after heating, holding at temperature for a period of time, and then cooling to the normal temperature after this (these relations were investigated in previous studies [7-9]), and
4. Finite Element Analysis (FEA) of the reconstruction of the behaviour of the car body side member under the three consecutive thermo-mechanical operations, dynamic plastic compressed deformation, restoration by heating, and then dynamic compressed deformation again. In these analyses, every subsequent FEA iteration must use the output data from the previous calculation as input data for next operation.

3. Methods of experimental studies

The mechanical properties of DP 780 steel samples were determined on a universal tensile testing machine. The chemical composition of the investigated steel DP780 is: 0.18 C; 0.5 Si; 2.0 Mn; 0.025 P; 0.8 Cr; 0.015 S.

Table 1. Mechanical properties of steel.

Material	Yield strength, MPa	Tensile strength, MPa	Maximum elongation, %
DP780	490	780	14

Monotonic tension and reverse loading (compression-tension) was executed on a universal tensile testing machine MTS 810, while fixing the sample to prevent buckling under compression and heating during the test. Elongation was recorded by laser extensometer LE-05. To prevent loss of stability of the test sample, it was supported by applying 3.35 kN lateral force by applying pneumatic cylinders on both sides. The stabilizing lateral force was applied to the sample through a set of rollers and Teflon coated plates in order to reduce friction.

When using the stabilizing lateral force it is necessary to perform a correction of output readings, taking into account the friction forces between the sample and the support plate and complex stress status of the test sample [10]. The coefficient of friction was determined by dividing the axial force during the normal loading. The average value in the series of similar experiments was 0.165.

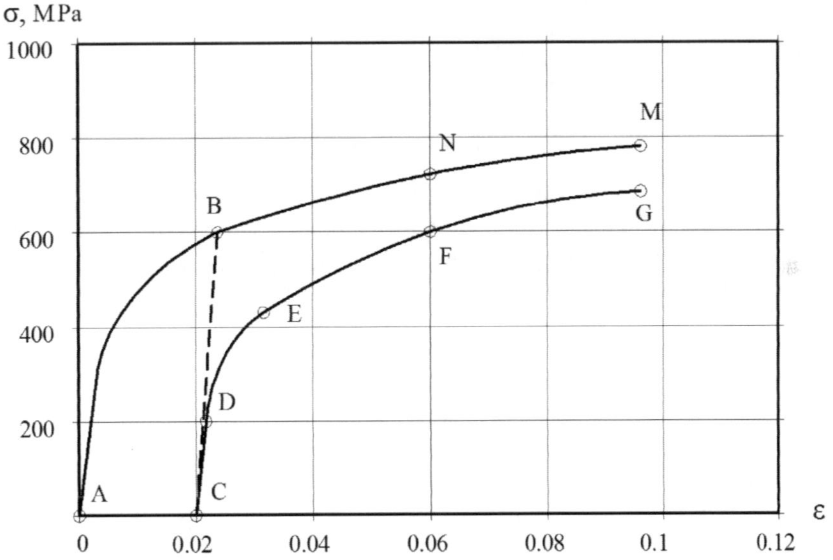

Figure 1. Compression-tension of DP780 dual-phase steel.

Specimen heating occurs as a result of heat exchange with the heating element via a plate with a Teflon coating. Oak spacers were used to isolate all the mechanisms of the testing machine. Temperature control was ensured by means of a calibrated thermocouple installed in the centre of the heating clamping plates. During stretching, the pre-compressed metal is heated to 150 °C, 300 °C, and 450 °C. These are the same conditions that the vehicle body undergoes during heated drawing.

The true behaviour of dual-phase steels is shown in figure 1. Samples of steel were pre-compressed to a value of effective residual strain of 0.02 (section A-B), the load was removed (section

B-C), and then the samples were stretched (section CG). Compression and tension are shown in the first positive quadrant of the graph (figure 1). While analysing the behaviour of the DP 780 dual-phase steel (figure 1), the following states can be singled out: the Bauschinger effect (point D), transition near the onset of hardening plasticity (point E), and softening in the area (F-G). In order to construct a model of the behaviour of such steel it is proposed to use a two-dimensional rheological Chaboche theory [4].

4. Main relations of the model accounting for technological material memory

4.1. Two surface plasticity model

According to theory, it is assumed that the stress tensor can be used to define both a surface of elasticity (limit of proportionality), f, and a surface of plasticity (yield strength of the material of the surface), F. Inside surface f the material behaves linearly elastic and obeys Hooke's law. Outside of F the material behaves plastically. The distance between the surface of plasticity and the surface of the elasticity is a continuous function, which describes the type of non-linear elastic deformation, *i.e.* variable elastic modulus.

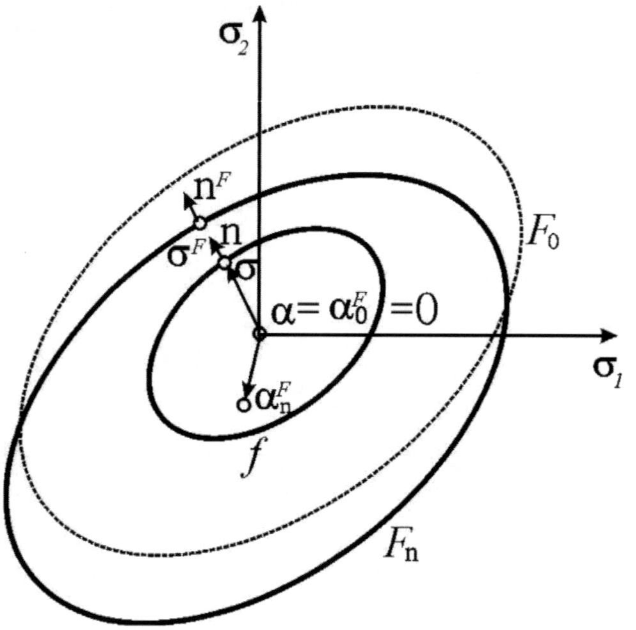

Figure 2. Graphic interpretation of the two-dimensional plasticity model.

Figure 2 shows the behaviour of the surface f, which represents the elastic state, and F, which determines the plastic state in the plane of the principal components of the stress tensor (for the two-dimensional case).

Surfaces f and F can be described by the equations:

$$f = \varphi(\boldsymbol{\sigma} - \boldsymbol{\alpha}) - r(\varepsilon^P, \dot{\varepsilon}^P, T) = 0, \qquad (1)$$

$$F = \Phi(\boldsymbol{\sigma}^F - \boldsymbol{\alpha}^F) - R(\varepsilon^P, \dot{\varepsilon}^P, T) = 0, \qquad (2)$$

where r and R limit the dimensions of surfaces f and F respectively, and the location of their centres is determined by the tensors of residual microstrain $\boldsymbol{\alpha}$ and $\boldsymbol{\alpha}^F$ respectively. The dimensions and coordinates of the centres of surfaces f and F do not change during the elastic deformation.

When stress tensor $\boldsymbol{\sigma}$ reaches a boundary of the surface f, the centre of the surface, $\boldsymbol{\alpha}$, begins to move along the surface in the direction of stress growth. When the inner surface, f, reaches the boundaries of the surface F, the material is then in the plastic state. According to the conditions of the surface translations, the contact occurs at the point congruent to the stress tensor. The increment of the stress tensor is directed outward from the surface f, $d\boldsymbol{\sigma} :: d\mathbf{n} > 0$. Both types of deformation, described by this theory – plastic deformation $\boldsymbol{\varepsilon}^P$ and elastic deformation $\boldsymbol{\varepsilon}^e$ – are present at this moment.

During deformation, the dimensions of the surfaces f and F may be changed to reflect the isotropic hardening of the material [4].

The Bauschinger kinematic hardening effect is simulated by the residual microstrain tensor. Both kinematic and isotropic hardening is pronounced in modern automotive steels, reflecting the capacity for significant energy absorption during deformation. The tensor of residual microstrain $\boldsymbol{\alpha}^F$ is described in accordance with the Chaboche model:

$$\boldsymbol{\alpha}^F = \boldsymbol{\alpha}_1^F + \boldsymbol{\alpha}_2^F \tag{3}$$

$$d\boldsymbol{\alpha}_1^F = \frac{2}{3} C_1 d\boldsymbol{\varepsilon}^P - \gamma \boldsymbol{\alpha}_1^F d\bar{\boldsymbol{\varepsilon}}^P \tag{4}$$

$$d\boldsymbol{\alpha}_2^F = \frac{2}{3} C_2 d\boldsymbol{\varepsilon}^P \tag{5}$$

The term, $d\bar{\boldsymbol{\varepsilon}}^P$, here is the equivalent increment of plastic strain

$$d\bar{\boldsymbol{\varepsilon}}^P = \left(\frac{2}{3} d\boldsymbol{\varepsilon}^P : d\boldsymbol{\varepsilon}^P \right)^{1/2} \tag{6}$$

Evaluation of the size of the surfaces of elasticity and plasticity is a major theoretical problem in the phenomenological approach, as it determines the start of the plasticity and residual deformation. The dimensions of the surfaces r and R are determined experimentally based on the elongation diagram of the metal. Evaluation of mechanical properties will lead to changes in size of R surface.

4.2. Empirical study of the influence of temperature on pre-compressed metal

The feasibility of repair of body parts has not yet been the subject of significant research, nor has the impact of local changes in strength, as a result of local heating, on the behaviour of the entire structure. Experimental studies to simulate the process of heated drawing of the deformed parts (including pre-compression and following drawing with heating) were carried out. Such a loading sequence is typical during the repair of a deformed vehicle.

Compression-tension diagrams were recorded during testing of the DP780 dual-phase steel (figure 1). A decrease in strength, ductility and a change in hardening rate, with increasing temperature, is demonstrated.

Empirical dependency is used to describe the behaviour of steel in the heated state. Dependency is represented as a result of multiplying the isotropic hardening h_1 and the heat evolution h_2.

$$\sigma = \sigma(\varepsilon, T_H) = h_1(\varepsilon, T_H) \cdot h_2(T_H) \tag{7}$$

The function h_1 reflects the impact of plastic deformation degree, taking into account the nature of hardening change under the influence of temperature. The term h_1 is a combined function of $g_1(\varepsilon)$ and $g_2(\varepsilon)$.

$$h_1(\varepsilon, T_H) = \beta(T_H)g_1 + (1-\beta(T_H))g_2 \, , \tag{8}$$

where T_H – heating temperature of steel; ε – accumulated plastic deformation of the metal. Function $g_1(\varepsilon)$ reflects the behaviour of the hardening curve at a temperature of 20 °C, while under the influence of heating, the function of ductility takes the form of $g_2(\varepsilon)$. The transition from one form to another is performed in accordance with the function $\beta(T_H)$. The function $\beta(T_H)$ changes from 0, at 20 °C, to 1, at 450 °C. For $\beta(T_H)$=0, h_1 degenerates to $g_2(\varepsilon)$, and for $\beta(T_H)$. = 1 – to $g_1(\varepsilon)$. The Zener-Hollomon parameter [11] was used to describe the work hardening of metal in normal conditions $g_1(\varepsilon)$ while Voce's function [12] was utilized to describe heated metal $g_2(\varepsilon)$.

$$g_1(\varepsilon) = K \cdot \varepsilon^n \, , \tag{9}$$

$$g_2(\varepsilon) = \sigma_0(1 - A\exp(-B\varepsilon)) \, . \tag{10}$$

We can transform (8) into

$$h_1(\varepsilon, T_H) = \beta(T_H)K \cdot (\varepsilon)^n + (1-\beta(T_H))\sigma_0(1 - A \cdot e^{-B\varepsilon}) \, . \tag{11}$$

For the $\beta(T)$ function it is convenient to use simple linear law:

$$\beta(T_H) = \beta_1 - \beta_2(T_H - T_0) \, , \tag{12}$$

where β_1, β_2 – constants derived from the experimental data; T_H – heating temperature of steel; and T_0 – base temperature (20^{0}C).

The need to develop this function arises from the diagram of dual-phase steel for stretching at various temperatures (Figure 3). The graph shows that the hardening rate depends on the temperature and decreases with temperature increase. This behaviour is not considered in the traditional phenomenological models.

A simple linear function will be used for the function $h_2(T_H)$. Heating during the repair cannot be controlled with high precision because of the large volumes of heated metal, the thermal conductivity and the specific heat of the material. Therefore, during the repair, the accuracy of linear dependence (even with the certain error) will be higher than the accuracy of the temperature control methods:

$$h_2(T) = \sigma(1 - \beta_3(T_H - T_0)) \, , \tag{13}$$

where β_3 – experimental constant.

The final iteration of the plasticity evolution equation for dual-phase steel, taking into account the isotropic hardening and heating:

$$R(\varepsilon, T_H) = [\beta_1 - \beta_2(T_H - T_0)K \cdot (\varepsilon)^n + (1 - \beta_1 - \beta_2(T_H - T_0))\sigma_0(1 - A \cdot e^{-B\varepsilon})][1 - \beta_3(T_H - T_0)] \, . \tag{14}$$

The empirical coefficients of the model were obtained by the method of least squares and are presented in table 2.

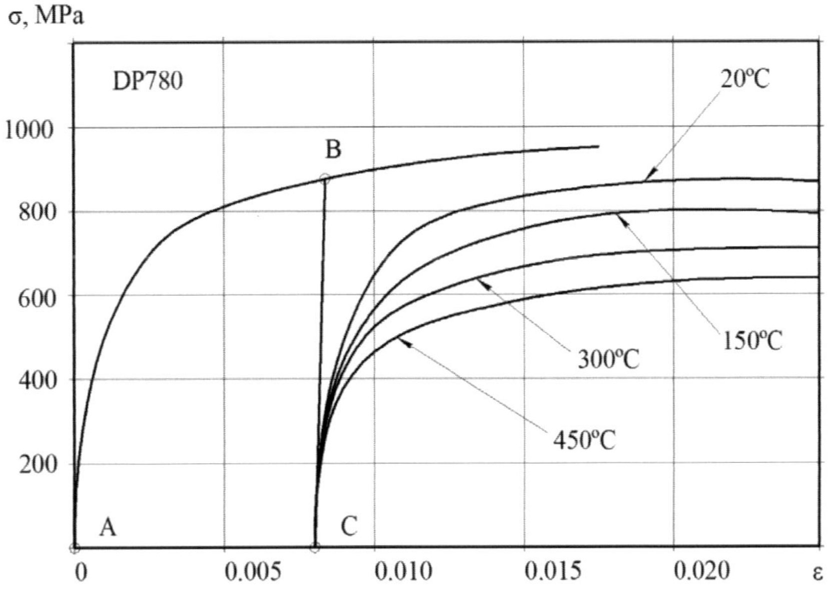

Figure 3. Compression-tension diagram of DP780 steel at 20 ^{0}C, 150 ^{0}C, 300 ^{0}C and 450 ^{0}C.

Table 2. Empirical model coefficients.

Steel	K, n	σ_0	A	B	β_1 β_2	β_3	R^2 $\sqrt{D}$, MPa
DP780	1655 0.213	752.1	0.265	30.31	0.507 0.00187	$5.8\cdot 10^{-4}$	0.998 2.5

4.3. Empirical study of the mechanical properties evolution after cooling, depending on heating temperature and hold duration

These studies were performed in previous investigations [7-9] and underpin the use of the final relations reflecting considerable degradation of the mechanical properties of dual phase steels following technological heating to the temperature more than 600 ^{0}C and for durations of more than 20 minutes. The lower temperatures and shorter durations did not affect mechanical properties so much; otherwise, the maximum temperature and hold duration, taking into consideration the simulation of the evolution of the plasticity limit during the repair, with non-isothermal heating, are proposed.

$$R(\varepsilon,T,t)=\left[(1-\alpha(T,t))K_2\cdot\varepsilon^{n_2}+\alpha(T,t)(\sigma_0-(\sigma_0-B_6)\exp(-a_1\cdot\varepsilon))\right]\cdot\left[1+\frac{C_1\cdot t^{n_5}\left(\dfrac{T}{C_3}+C_2\right)^{n_6}}{\sigma_0}\right], \tag{15}$$

where

$$\alpha(T,t)=\left(1-\frac{\sigma_{B0}-B_1\cdot t^{n_3}\left(\frac{T}{T_{AC1}}+B_2\right)^{n_4}}{\sigma_{B0}}\right), \tag{16}$$

T – heating temperature, ^{0}C; T_{AC1} – temperature of phase transition critical point AC_1; ε – accumulated plastic strain; σ_{BO} – tensile strength of the base metal MPa; t – holding time at the maximum temperature, s; $R(\varepsilon,T,t)$ – function of the limit of plasticity evolution after the cooling; and the following are empirical coefficients:

$K_2, n_2, \sigma_0, B_6, a_1, C_1, C_2, C_3, B_1, B_2, n_3, n_4$

Table 3. Empirical coefficients of evolution model of temperature degradation.

Steel	K_2, n_2	n_3, n_4, n_5, n_6	C_1, C_2	B_1, B_2, B_4, B_6	σ_0, σ_{BO}	a_1, T_{AC1}	R^2 $\sqrt{D}$, MPa
DP780	1655 0.213	0.66 6.5 0.66 3.44	2.38 0.4	0.034 0.024 2.21 300	260 780	8.56 724	0.999 0.7

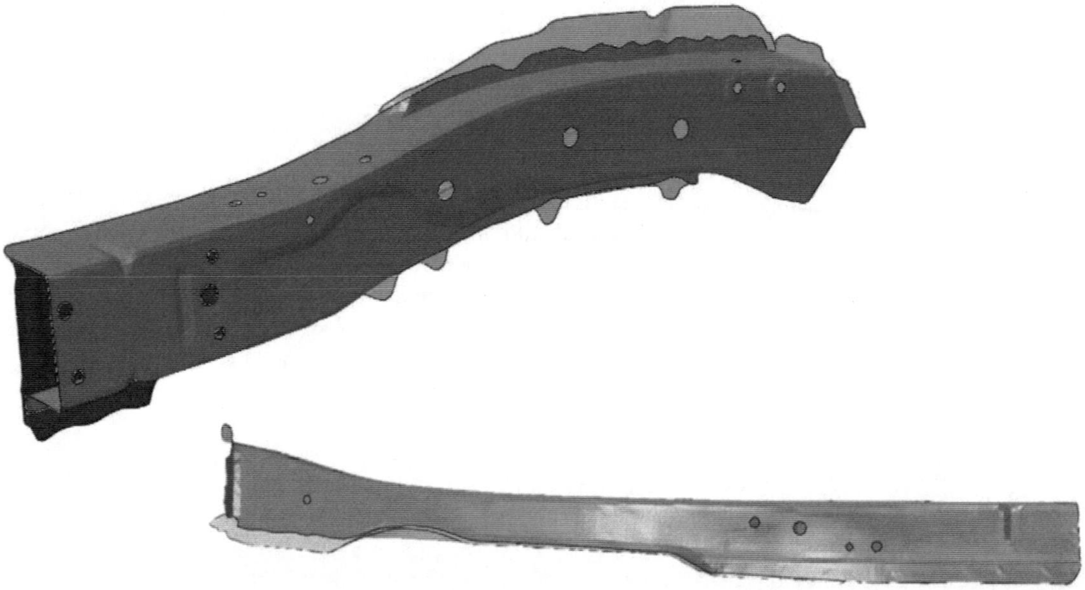

Figure 4. Shape of the side member under investigation.

5. FEA modelling of consecutive technological operations on side members

To evaluate the feasibility of dual-phase steel side member repair by drawing under heating, an computational experiment was conducted with a finite element model of the side member, using ABAQUS 6.11-3 Standard. Side member material properties were calculated in accordance with the UMAT procedure. To communicate with the ABAQUS pre-processor through the UMAT interface a program was developed that allows input of material properties with the help of the `INPUT FILE` file and `* USER MATERIAL` command.

Dual-phase steel DP780, with wall thickness of 1.5 mm, was chosen as the material of the detail. A curved shape was specified for the side member (figure 4), and was used to obtain strains that are typical for repairs using drawing techniques.

At the first stage of trial, the new side member was deformed by an impact with a flat rigid plate of mass 500 kg, travelling in axial direction at a speed of 16 m/s (figure 5).

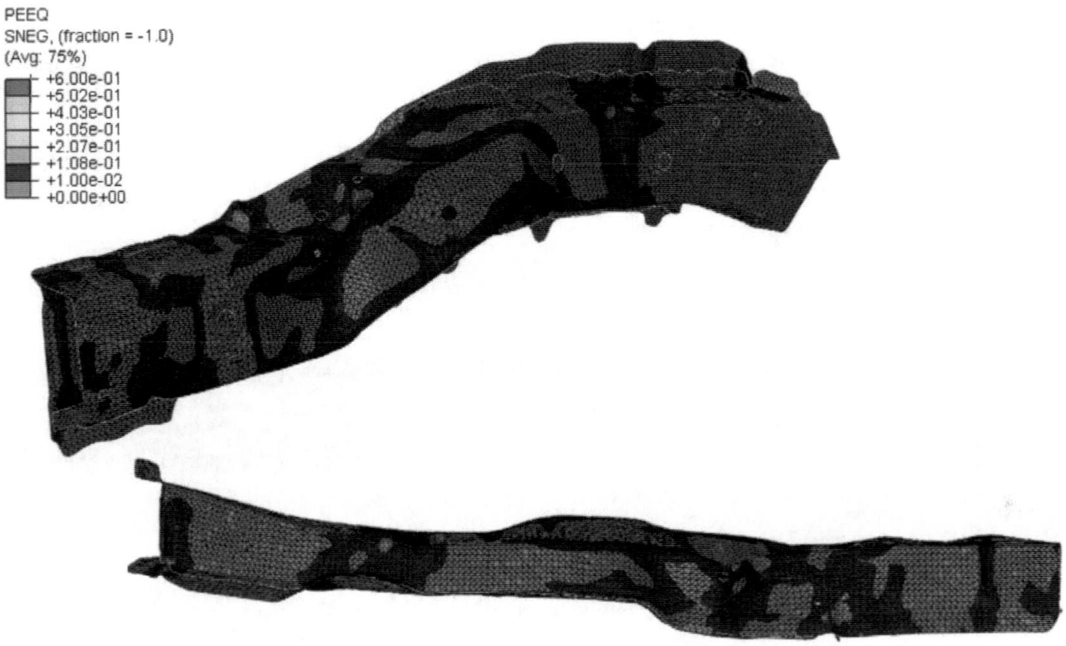

Figure 5. Deformation and stress in the side member after collision.

The analysis results of the new deformation of the side member (figure 5) show damage typical of frontal impact. The side member lost stability and bent at the base, where it is mounted, and at the tip, which is the point of force application. The most significant deformation occurred in the middle, as this is the site of application of the largest forces. The plastic strain energy in this case was equal to 10.5 kJ.

During the second phase of testing, the pre-deformed side member (figure 5) was subjected to a tensile load to simulate the process of repair. The data output file (Output Database) from ABAQUS has been exported to the original file (Input File), making it possible to transfer information about the accumulated strain in each element. Deformed elements were heated to reduce their limit of plasticity to the values of non-deformed metal in order to compensate for the strain hardening and equalize the strength characteristics of all elements. The heating temperature of each deformed element was calculated individually depending on the degree of deformation according to equation (14).

By substituting the value of the accumulated plastic deformation we determine the amount of heat required for each finite element. The same value of plasticity is achieved for all the elements of the detail by assigning the heating temperature to each of the elements.

Elements of the side member that used to have a strain now demonstrate a degradation of the ductility properties: this can be seen as a reduction of the hardening curve. The magnitude of degradation is in direct correlation with the degree of plastic deformation. Degradation is calculated according to equation (15).

The same distorted side member was subjected to drawing with overheating to investigate the effect of a non-optimal method of repair on the behaviour of steel. The side member was heated to a temperature of 700 °C. Optimally heated and overheated side members returned to the same shape after drawing, therefore completely restoring an original geometry.

The optimally heated and overheated repaired side members were deformed again at the next stage of the research. The numerical experimental conditions of load and speed were kept the same (figure 4). In the collision results (figure 6), both components exhibit an increase in deformation compared to a new side member. Note that all deformations arising from the three sequential numerical experiments was taken into account as part of the experimental methodology, therefore both the repaired parts have a significant number of elements which have modified properties, shown in the colour scheme of figure 5. The side member component repaired with minimal heating (figure 6 a) shows only a slight deviation of the shape compared to the new sample, with the strain concentration in the middle.

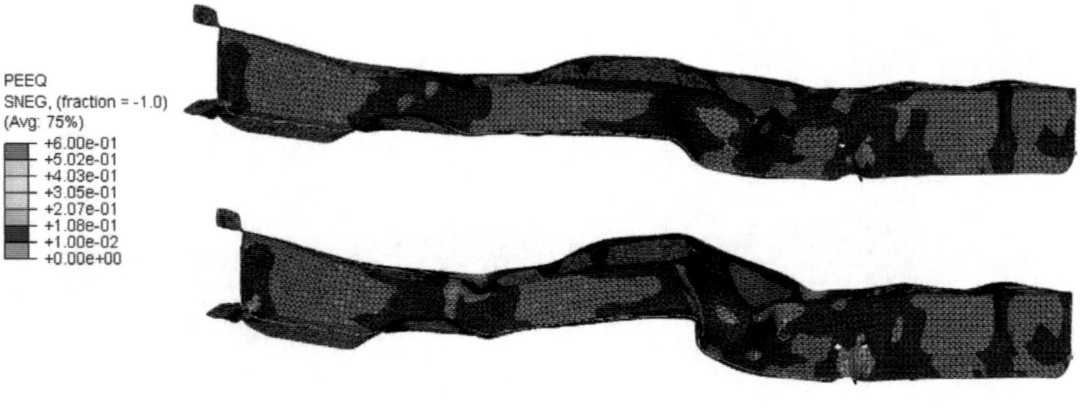

Figure 6. Deformation and stress in the repaired side members after repeated impact: (a) the optimally heated; (b) overheated.

6. Results and discussion

Simple visual evaluation (figure 6b) reveals differences in the shape of the overheated detail and demonstrates the presence of considerable strain along the length of the side member. The increased deformation of the part under load is presented in figure 7. The results for the model of the new side member reached the point of maximum deformation at 0.06 ms. Beyond that point it demonstrated some elastic unloading behaviour. The repaired, optimally heated side member finished the stage of energy absorption after 0.08 ms, while the repaired, overheated part continues deformation until past the 0.1 ms mark, showing no sign of elastic unloading. Comparison of the energy absorption graphs allows for a quantitative evaluation of changes in performance properties as a result of repair (figure 8).

The experiment conducted demonstrates that for a repair carried out with optimally heated drawing is able to sustain the level of passive safety of heat-treated steel car body in frontal impact. Energy absorption capability reduction of only 3% was observed in comparison with that of a new part. On the other hand, uncontrolled heating of the part during a repair can lead to a significant 18% drop in energy absorption and an increase in residual strain of 100%, which could be unacceptable for the purpose of providing body passive safety.

7. Conclusion

This series of studies has verified a methodology for the restoration of the material properties of passive safety components manufactured from heat-treated steels. Recovery technology and temperature conditions should be developed taking into account the evolution equations (14) and (15). The heating temperature must be strictly monitored during the repair, in accordance with the designed conditions. Adherence to the method of recovery could enable a significant reduction in body repair cost by allowing restoration of parts that are currently being replaced.

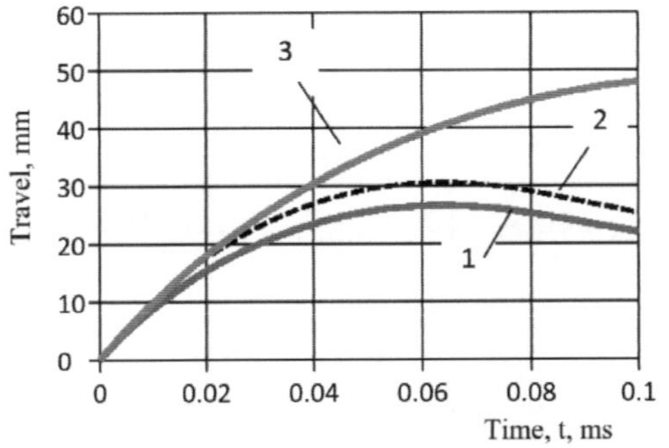

Figure 7. Side member travel during impact.

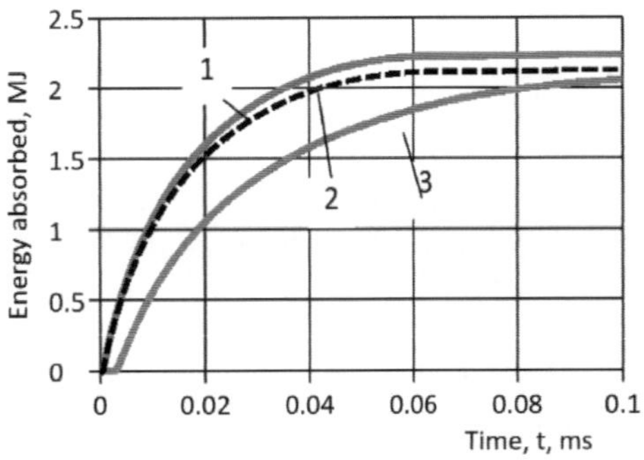

Figure 8. Collision energy absorption.

1 – a new side member; 2 – optimally heated during the repair; 3 – overheated during the repair.

The investigation confirms that the algorithm developed for the calculation and modelling of mechanical properties, which are then used in consecutive FEA iterations, can be used to predict the long term life behaviour of engineering machine components. The numerical calculation procedure consists of repeated numerical simulations of real technological operations on a real component. Such an approach is time resuming, but enables the acquisition of numerical results showing mechanical properties that reflect real anisotropic, and heterogeneous distortion, behaviour. Additionally this provides the opportunity for optimisation of future planned technological operations, such as the temperature control performed during this experimental study.

Acknowledgement

The research was funded by the Ministry of Education and Science of the Russian Federation on the basic part of the state assignment of the Sevastopol State University №2014 / 702 (Project 3867).

References

[1] Falaleev AP, Meshkov VV, Vetrogon AA and Shymchenko AV 2016 "Evolution of mechanical properties of boron/manganese 22MnB5 steel under magnetic pulse influences" *IOP Conf. Ser.: Mater. Sci. Eng.* **110** 012107

[2] Marcadet SJ and Mohr D 2015 "Effect of compression-tension loading reversal on the strain to fracture of dual phase steel sheets" *International Journal of Plasticity* **72** 21-43
[3] Barlat F, Vincze G, Grácio JJ, Lee MG, Rauch EF and Tomé CN 2014 "Enhancements of homogenous anisotropic hardening model and application to mild and dual-phase steels" *International Journal of Plasticity* **58** 201-18
[4] Chaboche JL 1989 "Constitutive-Equations for Cyclic Plasticity and Cyclic Viscoplasticity" *International Journal of Plasticity* **5** 247-302
[5] Zang SL and Guo C 2007 "A constitutive model for spring-back prediction in which the change of Young's modulus with plastic deformation is considered" *International Journal of Machine Tools and Manufacturing* **47** 1791-7
[6] Eggertsen PA and Mattiassou K 2009 "On the modeling of the bending-unbending behavior for accurate springback predictions" *International Journal of Mechanical Sciences* **51** 547-63
[7] Фалалеев АП 2012 "Влияние ремонтних нагревов на свойства автомобильной двухфазной стали" *Вісник СевНТУ* **133** 174-8 [Falaleev AP 2012 "Influence of repairs with heating on the properties of automotive dual-phase steel" *Visnik SevNTU* **133** 174-8 (in Russian)]
[8] Фалалеев АП 2012 "Влияние температурных режимов кузовного ремонта на эволюцию механических свойств двухфазных автомобильных сталей" *Вісті Автомобільно-дорожнього інституту ДонНТУ* **12** 64-7 [Falaleev AP 2012 "Influence of temperature modes of body repair on the evolution of the mechanical properties of automotive two-phase steels" *News of Automobile and Road Institute National Technical University* **12** 64-7 (in Russian)]
[9] Фалалеев АП 2012 "Влияние температуры и пластической деформации на эволюцию механических свойств двухфазных автомобильных сталей" *Вісник Чернігівського державного технологічного університету. Серія "технічні науки"* **51**(4) 104-7 ISSN 22257551 [Falaleev AP 2012 "Influence of temperature and plastic deformation on the evolution of the mechanical properties of automotive two-phase steels" *Herald Chernihiv State Technological University. A series of "Technical Science"* **51**(4) 104-7]
[10] Boger RK, Wagoner RH, Barlat F, Lee MG and Chung K 2005 "Continuous, large strain, tension/compression testing of sheet material" *International Journal of Plasticity* **21** 2319-43
[11] Hollomon JH 1945 "Tensile deformation" *Transactions of AIME* **162** 268-90
[12] Voce E 1948 "The relationship between stress and strain for homogeneous deformation" *Journal of the Institute Metals* **74** 537-62

IOP Conf. Series: Materials Science and Engineering **153** (2016) 012012 doi:10.1088/1757-899X/153/1/012012

Study variants of hard CFRP reflector for intersatellite communication

PV Prosuntsov, SV Reznik, KV Mikhailovsky, AD Novikov, Zaw Ye Aung

Rocket and Space Composite Structure Department Bauman MSTU,
5/1, 2nd Baumanskaya str., Moscow, 105005, Russia

E-mail: sergeyreznik@bmstu.ru

Abstract. The paper deals with the justification of space antennas reflector layout for advanced telecommunication satellites. The selection of design decisions is based on numerical simulations of heat transfer and mechanics processes characteristic of the geostationary orbit conditions. The advantages of parabolic shell of small thickness reflector scheme reinforced with star-shaped ribs on the convex side are demonstrated.

1. Introduction

Parabolic reflectors of mirror space antennas are widely used in space communication systems. Small linear density and high dimensional stability are the most important operating characteristics. Increasing the operation frequency of antenna increases the accuracy requirements of the reflector. Shape tolerances of the reflector must not exceed the $\Delta = \Lambda/16$ value, but tighter tolerances $\Delta = \Lambda/50$ can be used, where Λ is the antenna operating wavelength [1].

Carbon fibre reinforced plastics are commonly used as a material for reflectors due to the high specific strength, low coefficient of thermal expansion (CLTE) and a relatively high thermal conductivity.

2. The current state of research in the field of space mirror antennas reflector

Since the beginning of the 1990s, three-layer membrane layouts consisting of carrier layers, honeycomb core and elements of the framework have been common. An example of such design is the ETS-VI satellite, developed by Toshiba, Japan [2]. The folding reflector segments are sandwich panels with a honeycomb core of aluminum and carrier layers made of carbon fiber reinforced plastic (CFRP) [3]. Similar reflectors have been created by HPS GmbH, Germany (figure. 1) [4]. This type of reflector has high stiffness parameters but rather poor weight characteristics, in the order of 2.5 kg/m^2.

An alternative to the three-layer membrane layout is a dual gridded reflector made of carbon plastic (figure 2) [5]; however, this design also leads to increased structural weight. The diameter of the HPS GmbH reflector is 1.2 m, the reception frequency of the antenna is 27.5-30.0 GHz, and the transmission frequency is 17.5-20.5 GHz. The linear density is 3.09 kg/m^2. The reflector linear density can be reduced using smooth thin-walled shells with different kinds of reinforcement structures (rib layout). The ultralight carbon fibre reflector designed by ESA for the Artes 3 and Artes 4 missions is an example of such a type structure. The diameter of the Ka-band antenna reflector is 2.6 m [6].

Figure 1. Three-layer membrane scheme created by HPS GmbH.

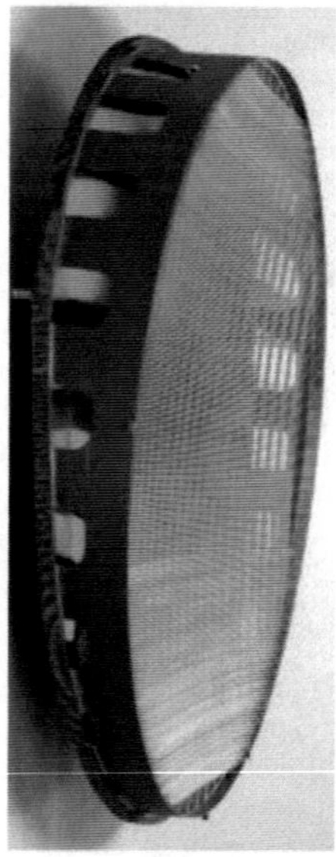

Figure 2. Dual gridded reflector by HPS GmbH.

Another design variant is that of a smooth shell with ribs, as in the Artemis satellite. The reinforcement element was integral with the shell of the reflector (figure 3), with a diameter of 2.85 m, and Ka- and S-band of the antenna [7]. The Eutelsat 115 West B satellite mirror space antennas reflectors are rotational shells with a 3D reinforcement structure on the backside. A feature of the reinforcement construction is that it is a hoop-like structure. The antenna operates in Ku-band and the diameter of the reflecting surface is about 2 m [8].

Space System Loral, USA, designed the Intelsat 30 satellite where Ku-band antennas where 2 meter diameter reflectors are installed [9-11]. A truss connected to the shell is employed as a reinforcement. ISS Reshetnev, Russia, jointly with Thaled Alenia Space, Italy-France, developed a Russian communication satellite, Yamal-402, with a rigid ultralight space antennas reflector. A 3D truss rigidly fixed to the reflector serves as a reinforcement for the 1.75-meter Ku-band reflector (figure 4) [12, 13]. For this type of reflector, the linear density characteristic is in the order of 2 kg/m^2, but to ensure the rigidity of the construction it is necessary to justify the selected layout of ribs. There is currently no theoretical justification for this selection.

3. Reflector structural design selection

The structural design of the reflector means the totality of its basic parameters determining the shape and size of the reflector, manufacturing principles and the materials used. Analysis of domestic and foreign experience has shown that creation of ultralight reflector based on thin ribbed shell of CFRP is the most advanced solution. Using this layout reduces the linear density compared to conventional three-layer constructions substantially, by several times.

A quasi-isotropic multilayer structure ensures the lowest value of reflector shell CLTE. It also provides the highest value and uniform thermal conduction coefficient. In this research, a layup of

[0°/+30°/–30°/+60°/–60°/+90°] Aspro A-50 fabric was used (layer thickness – 0.1 mm). The reflector linear density was 1.15 kg/m^2. The data were used for the design of the reflector are listed in table 1.

Figure 3. Artemis satellite.

Figure 4. Yamal-402 satellite antenna reflector.

Table 1. Data used for the design of the reflector.

Reflector dimension	Value, mm
Diameter (aperture)	1200
Focal length	500
Construction depth	180

The rib layout is selected based on comparative analysis of temperature and stress-strain state of the reflector with various reinforcement patterns. Four candidate reinforcement patterns were selected after considering Russian and foreign experience (figure 5):

a) A 30 mm height five-pointed star pattern. The rim of the reflector is supported by a circular rib 30 mm high.
b) A constant rib height isogrid pattern where the isogrid triangle finning is positioned at the back surface of the reflector. The step and height of ribs is 125 mm and 20 mm respectively. The rim of the reflector is supported by a circular rib 20 mm high.
c) A variable rib height pattern similar in type to the previous variant, but with non-constant rib height from 30 mm (at the edge) to 3 mm (in the centre). The rim of the reflector is supported by a circular rib 30 mm high.
d) A 30 mm height six-pointed star pattern. The rim of the reflector is supported by a circular rib 30 mm high.

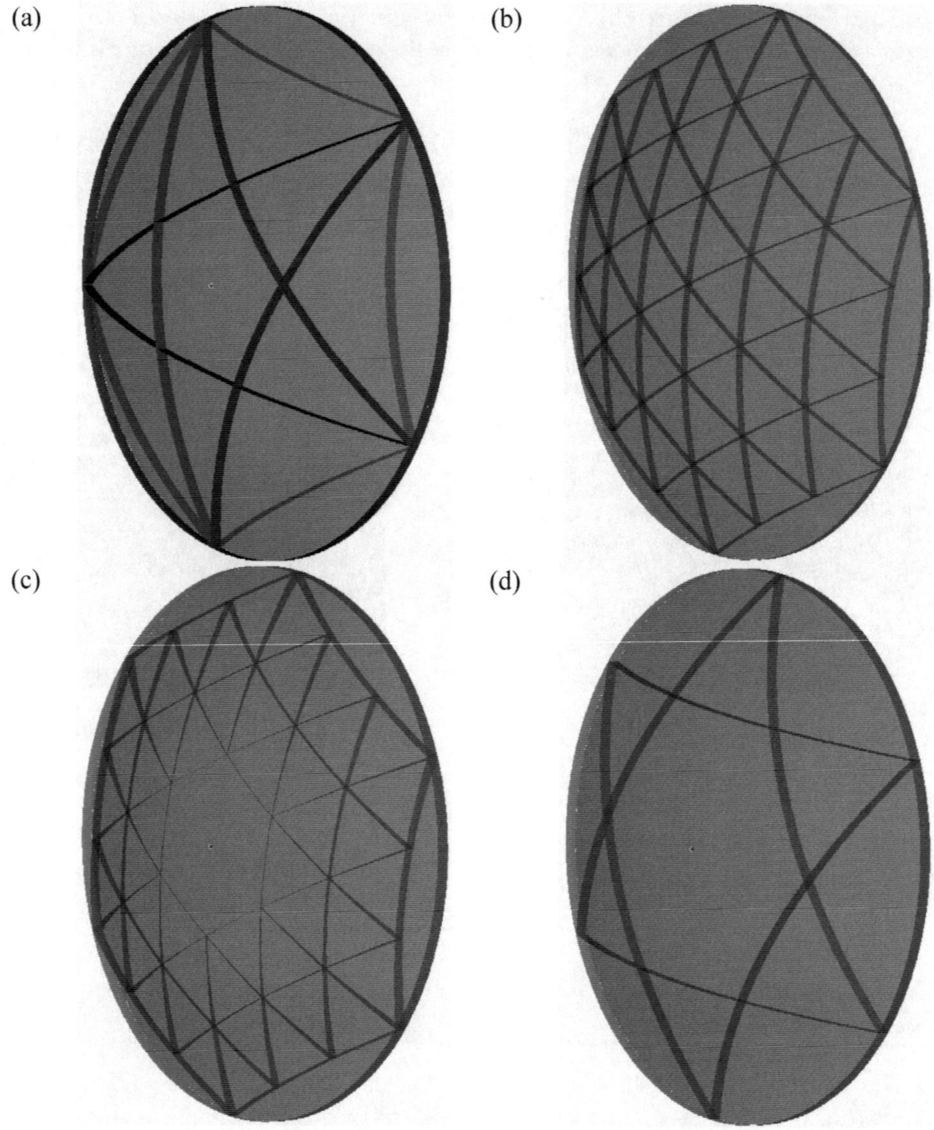

Figure 5. Variants of reflector layouts: a) five-pointed star, b) isogrid structure, constant rib height, c) isogrid structure, variable rib height, d) six-pointed star.

4. Reflector temperature and stress-strain state simulation

The flight of the spacecraft with a reflector on a geostationary orbit at the March equinox was considered. The rotation axis of the parabolic reflector shell would be directed toward the Earth and its transverse axis would coincide with the spacecraft velocity vector. In determining the structure thermal deformation it was assumed that the reflector was rigidly fixed via the adapter to its outer edge.

Incident solar and reflected radiation fluxes and the Earth thermal radiation were defined for the temperature state simulation, using the finite element analysis Siemens Femap 11.1 package [14]. The

IOP Conf. Series: Materials Science and Engineering **153** (2016) 012012 doi:10.1088/1757-899X/153/1/012012

reflector temperature field simulation results, also calculated using the finite element analysis (FEA) Siemens Femap 11.1 package, showed that the highest temperature difference and the thermal deformation in the reflector structure arise at the moment of its turning around the axis relative to the sun, which corresponds to 25200 seconds flight (figure 6, 7). Reflector finite element models include more than 5600 2D elements.

5. Results analysis

The thermal deformation analysis of the reflector (table 2) showed that, with almost identical values of the mass and linear density, the lowest temperature deformation is given by the five- and six- pointed star-shaped reinforcement pattern.

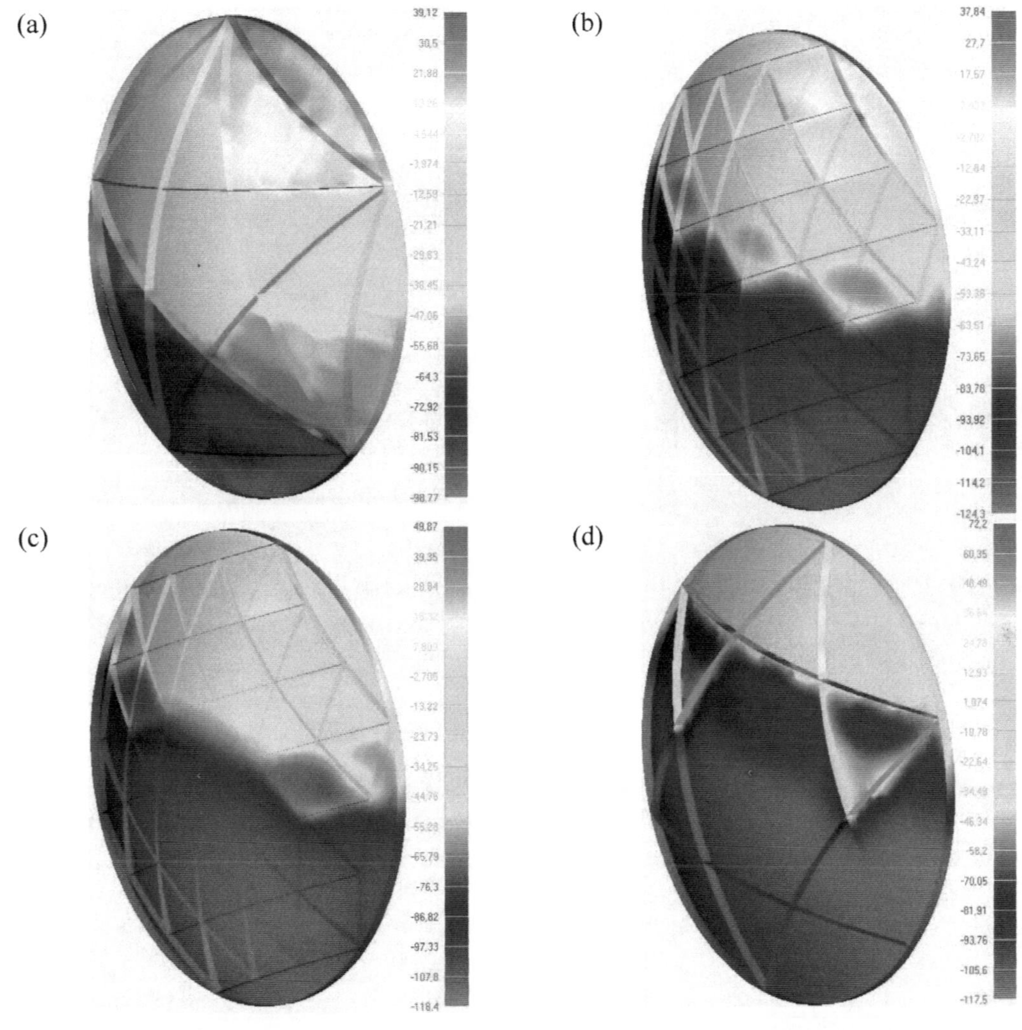

Figure 6. The temperature distribution on the surface of the reflector at 25200 s moment for different reinforcement patterns: a) five-pointed star, b) isogrid structure, constant rib height; c) isogrid structure, variable rib height; d) six-pointed star

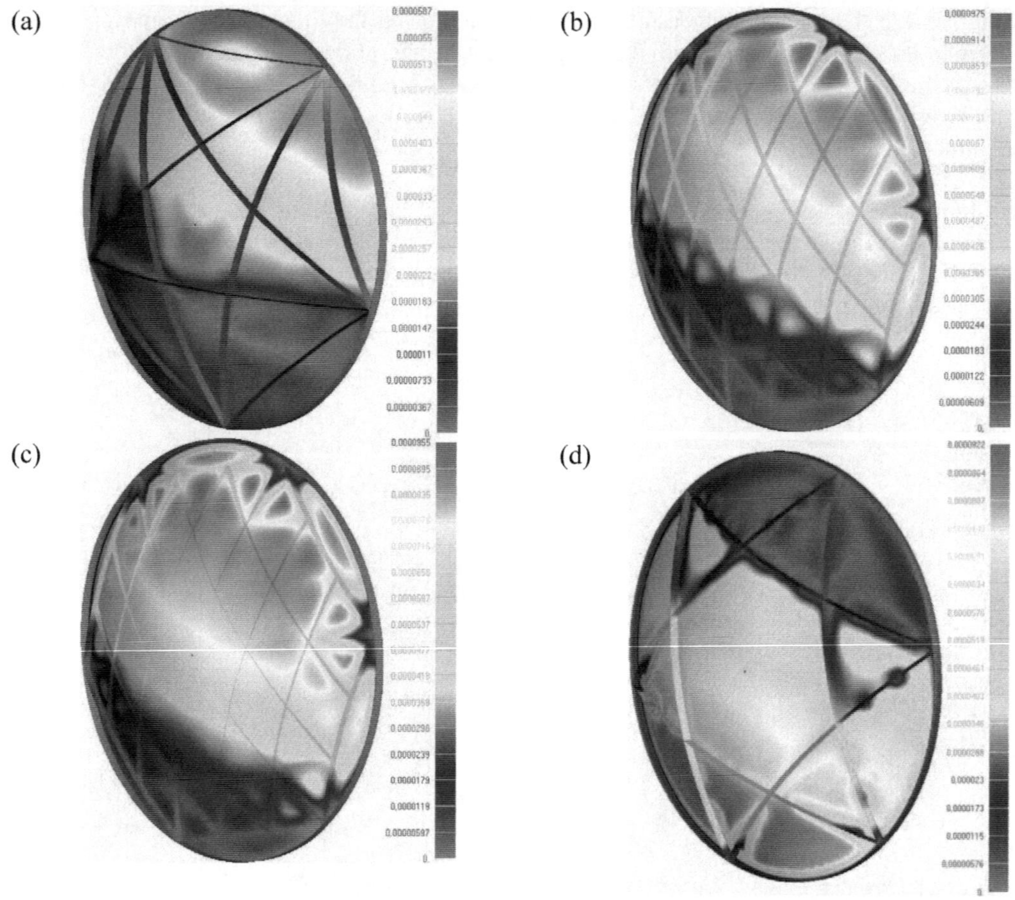

Figure 7. The thermal deformations on the surface of the reflector at the time 25200 s moment for different reinforcement patterns: a) five-pointed star, b) isogrid structure, constant rib height; c) isogrid structure, variable rib height; d) six-pointed star

Table 2. Comparison of weight and temperature values for different reflector reinforcement patterns

Reinforcement pattern	Mass, kg	The linear density of the reflector, kg/m^2	Maximum temperature deformation, mm
Five-pointed star	1.502	1.224	0.0587
Isogrid structure, constant rib height	1.549	1.262	0.0975
Isogrid structure, variable rib height	1.479	1.205	0.0955
Six-pointed star	1.513	1.235	0.0922

6. Discussion

Not only linear density and temperature deformations are important for space reflectors. As is shown by the analysis, the isogrid structure with the variable rib height has the lowest value of linear density; however, this layout is rather hard to manufacture. The lower are the ribs in the pattern, the easier it is in fabrication. First of all, the number of ribs is important in terms of the number of ribs joints. That is what determines the manufacturing precision. There are significantly fewer ribs joints in the five-pointed and six-pointed star patterns. This was the reason for choosing these patterns for the further analysis.

7. Conclusions

Four perspective reinforcement patterns of superlight Q/V-band 1200 mm diameter reflector were examined. Thermal and stress-strain state was obtained. It is demonstrated that the lowest value of deformation (less than 0.06 mm) was obtained for the five-pointed-shaped shell.

Acknowledgments

Some results of this work were obtained in the framework of the project №1864 under the state commission №2014/104 for the public research activity as part of the commission by the Ministry of education and science of the Russian Federation.

References

[1] Watanabe M, Misawa M, Minomo M, Yasaka T 1982 *High accuracy deployable antenna for communications satellites* (Proc. 13-th Int. Symp. On Space Technol. and Sci. – Tokyo, Japan)
[2] Goodwin J, Nevarez R, Funakawa K, Norii M 1986 *Engineering test satellite VI* (Japan: National Space Development Agency, Japan)
[3] Pfeiffer K E 2007 *Highly stable CFPR antenna reflectors* (Germany: HPS GmbH)
[4] Imbriale W A 2006 *Spaceborne antennas for planetary exploration* (USA: Jet Propulsion Laboratory California Institute of Technology)
[5] 2014 *New concepts for dual gridded reflectors* (Germany: HPS GmbH) http://www.hps-gmbh.com/projects-descriptions-mainmenu-79/lightweight-structures-mainmenu-51/23-kadgr.html
[6] 2014 *Large deployable ultra-light reflector for Ka-band applications* (ESA) https://artes.esa.int/projects/large-deployable-ultra-light-reflector-ka-band-applications
[7] 2009 *Advanced Data Relay and Technology Mission Satellite* (Weebau) http://weebau.com/satellite/A/artemis.htm
[8] 2015 *EUTELSAT 115 West B satellite gears up for launch* (Eutelsat) http://news.eutelsat.com/pressreleases/eutelsat-115-west-b-satellite-gears-up-for-launch-1122494
[9] 2014 *Intelsat 30,31/DLA 1,2* (Gunter's Space Page) http://space.skyrocket.de/doc_sdat/intelsat-30.htm
[10] 2014 *Encapsulating Intelsat 30/DLA-1 for its journey into space* (Intelsat) http://www.intelsat.com/blog/intelsat-launches-blog/encapsulating-intelsat-30dla-1-for-its-journey-into-space
[11] 2014 *SSL-built Asiasat 6 begins post-launch maneuvers according to plan* (SSL) http://sslmda.com/html/pressreleases/pr20140908.html
[12] 2013 *Technical Characteristics of Yamal-402 sattelite* (Gazprom Space Systems) http://www.gascom.ru/ru/new_projects/yamal_402/technical_performance/index.php
[13] 2012 *Nouvel echec pour un lanceur russe Proton* (Futura Espace) http://www.futura-sciences.com/magazines/espace/infos/actu/d/astronautique-nouvel-echec-lanceur-russe-proton-43298/
[14] 2014 *Femap user guide* (Siemens PLM) https://extras.springer.com/2002/978-3-642-62564-0/Wtp2000/Femap/docs/user.pdf

Optical characterisation of metallic meshes for space antennas transformable reflectors

RA Mironov, SV Reznik, RV Rukavishnikov, VA Shishulina and VA Zavaruev

Department Rocket and Space Composite Structure Bauman MSTU, 5/1, 2nd Baumanskaya str., Moscow, 105005, Russia

E-mail: sreznik@bmstu.ru

Abstract. The paper deals with metallic meshes for space antenna transformable reflectors. Optical characteristics of four metallic meshes (15 micron tungsten, 15 micron gold-plated tungsten, 20 micron molybdenum, and 50 micron Ni-coated steel) were investigated with the Nikolet IS50 spectrophotometer. The spectral optical characteristics of these metallic meshes in the range from 0.4 to 2.5 microns are required for thermal design of space antenna reflectors.

1. Introduction

The Creation of Large Deployable Antennas (LDA) has been one of the most perspective directions in the development of global telecommunications systems for already more than 40 years [1, 2]. Such companies as Harris Corporation and Astro Mesh (which is a subsidiary of Northrop Grumman Corporation) are designing and manufacturing space transformable reflectors with large diameters of up to 25 meters [3-5]. These structures generally have a membrane or an umbrella layout. Researches in the field of such antennas are conducted in Europe and Japan [6-10]. A number of projects have been performed in Russia by the Joint-Stock Company (JSC) “Special Design Bureau of Moscow Power Engineering Institute”, JSC “SP Korolev Rocket and Space Corporation “Energia”, JSC “Academician MF Reshetnev Information Satellite Systems” [11-15]. Features common for all existing LDA include metallic reflecting meshes, ropes of polymeric fibres and load-bearing elements made of polymer composite materials.

2. Relevance of the research

Increasingly high requirements are imposed on modern space antenna reflectors both in terms of durability and precision. In order to achieve minimal shape and size deviations (maximum value of $\Lambda/16$ where Λ is the antenna operating wavelength), the design process must attend to the properties of each and even the smallest element.

When in orbit the reflector of mirror space antenna will be in high vacuum and will be heated by the solar radiation thermal flux, the solar flux reflected from the Earth and the Earth's own radiation. The main source of the external thermal influence is the heat flux of direct solar radiation.

In geostationary orbit, twice a year, at around the vernal and autumnal equinoxes, the satellite will be in the Earth's shadow with the maximum duration of up to 71 minutes. Thin walled elements of the LDA reflectors can be cooled down to low temperatures (less than 100 K) when in shadow and can be heated intensively when in daylight sector of orbit [12]. Temperature dynamic causes temperature gradient, which could become a reason of structure deformations followed by angular pattern break down. Therefore, one of the main stages of the mesh reflector design is thermal regimes modelling. For the modelling of the thermal regimes, not only the reflector, but also of the whole telecommunication satellite, the following data is required: the quantity of solar flux absorbed into and transmitted through the mesh surface, the intrinsic temperature of mesh surface, and the amount of heat energy emitted to the external environment. The accuracy of modelling depends on the input data, accuracy of the thermal physical, and optical characteristics of the materials utilized.

Metal meshes are manufactured by weaving tungsten, molybdenum or steel wire with a diameter of 15 to 50 micrometers and have complex spatial structure. Metallic knitted mesh enables both shape stability at larger sizes and lower mass per unit length [14, 15].

The original production technologies for various knitted metallic meshes of radio-technical purpose are presented in the works [16-21]. The main emphasis of the mentioned papers is on getting meshes from various fibres with defined cell sizes and required rigidity. There is little information in literature concerning the characterisation of radio-technical metallic meshes for thermal design. There is also still a need for further research into the optical properties of metal meshes. Their characterization is complicated by the structure, in which voids account for more than 80%, and a need for characterization of the variability of the surface profile and the dependence of the characteristics on the magnitude and direction of the loads.

In paper [12], thermal design results for the prospective LDA, including the reflector fabricated of gold-plated tungsten mesh and the use of a multi-joint manipulator for moving the reflector to the operating position. Optical characteristics of the metallic mesh were obtained theoretically for the fibrous periodic structure model. Independent calculations of the thermal regimes were accomplished by the company Alenia Spazio with the help of ESATAN and ESARAD computer codes, developed by the European Space Agency (ESA) [12] and Bauman MSTU with the help of the Bauman MSTU in-house CAR computer code. The metallic mesh was modelled as a thin layer of semi-transparent material with the heat emitted from surface to the external environment. Results of experiments conducted in ESTEC (Noordwijk, the Netherlands) [12] have shown satisfactory fit with those of theoretical and experimental data.

In papers [22, 23] the method of experimental definition of optical characteristics of mesh made of 50 microns diameter nichrome wire is presented. Transmission and reflection coefficients were determined on the standard optical equipment with the original VL Churkin's devices [24] within the range of 0.2-10 μm length in Joint-Stock Company "Dolgoprudny Design Bureau of Automatics". The authors tested plate patterns of mesh in a vacuum chamber simulating single side solar heating [22, 23]. The temperature patterns were measured with high accuracy by using ultra small thermocouples with 30 μm diameter thermocouple wire. In accordance with the theoretical estimations, the temperature measurement systematic error did not exceed 1.5 K in a temperature level of 420 K [22, 23]. The relationship A_s/ε obtained as a result of the experimental data interpretation showed satisfactory agreement with the data obtained with standard optical equipment.

Paper [25] describes the method and results of the experimental research of mesh optical characteristics (absorption coefficient of solar emission A_s, coefficient of thermal emission ε, and transmission D_s, in the range of solar spectrum 0.2 – 2.5 μm wavelength and in the infrared region with $\Lambda \sim 10$ μm). Standard equipment such as laid-on photometer FM-59 and thermoradiometer TRM-I using laying of mesh on the base with the different coefficients A_s, ε, including the models of an ideal black body was used.

The mesh transmittivity indicatrix was obtained for different solar emission incident angles using an imitator of using an imitator of solar radiation and PCFS photo sensor. It was determined that transmission of any mesh type has a particular nature: the beam divergence angle does not exceed 15 degrees. The mesh and antenna elements optical characteristics measurements results were used for the analysis of the reflector thermal regimes under operation conditions. The papers referenced give an idea about the current research experience, but the appearance of new mesh materials inspires further updating of the methods.

3. Test samples and the test setup

Four types of metal mesh were selected for testing: 15 micron tungsten wire (figure 1), 15 micron gold-plated tungsten wire (figure 2), 20 micron molybdenum wire (figure 3), and 50 micron nickel-coated steel wire EI 708A (figure 4). Samples have different weaving patterns and cell size. The main characteristics of the meshes are shown in table 1.

Table 1. Mesh characteristics

Material	Coating	Wire diameter, μm	Cell size, mm
Tungsten	none	15	0.2×0.8
Tungsten	Gold	15	0.2×0.8
Molybdenum	none	20	0.2×0.8
Steel	Nickel	50	1.7×1.7

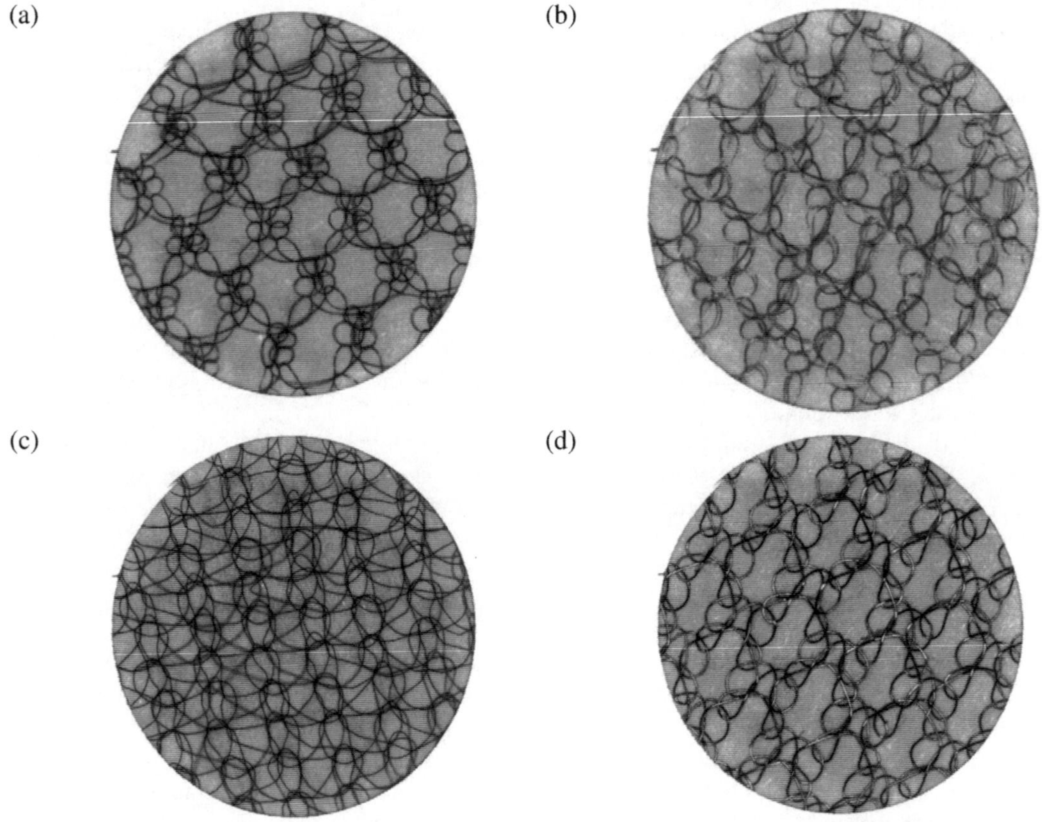

Figure 1. (a) Tungsten mesh microstructure, (b) Gold-plated tungsten mesh microstructure, (c) Molybdenum mesh microstructure, (d) Steel mesh microstructure.

An important stage of the experiment was to prepare frames for samples. It was necessary to take into account the test stand features, the size of the light beam passing through the samples, the uniformity of the sample material and their optimal tension. Metal mesh samples were cut from blanks and fixed in a cardboard frame with a 25 to 50 mm hole.

4. Experiment description

Reflection coefficients were determined using the near infrared range (NIR) spectroscopy by means of Nicolet iS50 FT-IR spectrometer (Figure 2) [26, 27].

Figure 2. Spectrometer Nicolet iS50 [26].

The physics of this method is that the source radiates infrared electro-magnetic waves, and the electro-magnetic radiation passes through Michelson interferometer and the sample. The detector registers the dependency of the output light flux intensity on the optical path difference in the different parts of the interferometer. The interferogram contains full information on the spectral nature of the radiation.

Integrating sphere is used as a measuring instrument. The integrating sphere method determines diffuse transmission and reflection in 0.2 – 2.6 μm spectral region. The integrating sphere is a hollow sphere; the inner surface is coated with a diffusely scattering material.

An integrating sphere is used as a measuring device. The diffuse transmittance and reflectance in 0.2 – 2.6 μm spectrum range can be determined with the integrating sphere. This main measuring element in photometers and spectrophotometers is a hollow ball with diffusing internal surface. The device was calibrated immediately before the experiment.

5. Results and discussions

The data obtained using Nicolet iS50 were processed by means of specialized software package OMNIC 9 resulting in spectral dependencies of reflectivity R_υ and transmittivity D_υ. Absorptivity A_υ values were calculated using the equation:

$$A_\upsilon = 1 - D_\upsilon - R_\upsilon.$$

Experimental results are shown in graphs as dependencies of spectral optical properties on the wavelength in the infrared and optical spectra (0.4 – 2.5 micron) at normal incidence (figures 3-6).

The absorptivity of all types of metal mesh did not exceed 0.2 in the whole spectral range. The transmittivity of all meshes, except the molybdenum mesh, decreased smoothly and very gradually with increasing wavelength; however, it still remained in the 0.75-0.85 range. The transmittivity of the molybdenum mesh increases slightly with wavelength in the infrared range, but does not exceed 0.89. The lowest absorptivity over the greatest part of the spectrum was demonstrated by the gold-plated tungsten mesh, which indicates an advantage for solar radiative heating at GEO.

Reflectivity of all samples except gold-plated tungsten mesh decrease smoothly with the wavelength increase. The greatest reflectivity values were registered for the nickel-coated steel mesh;

IOP Conf. Series: Materials Science and Engineering **153** (2016) 012013 doi:10.1088/1757-899X/153/1/012013

greater than 0.8. On the whole, the data obtained were in alignment with the previously obtained data for metal meshes.

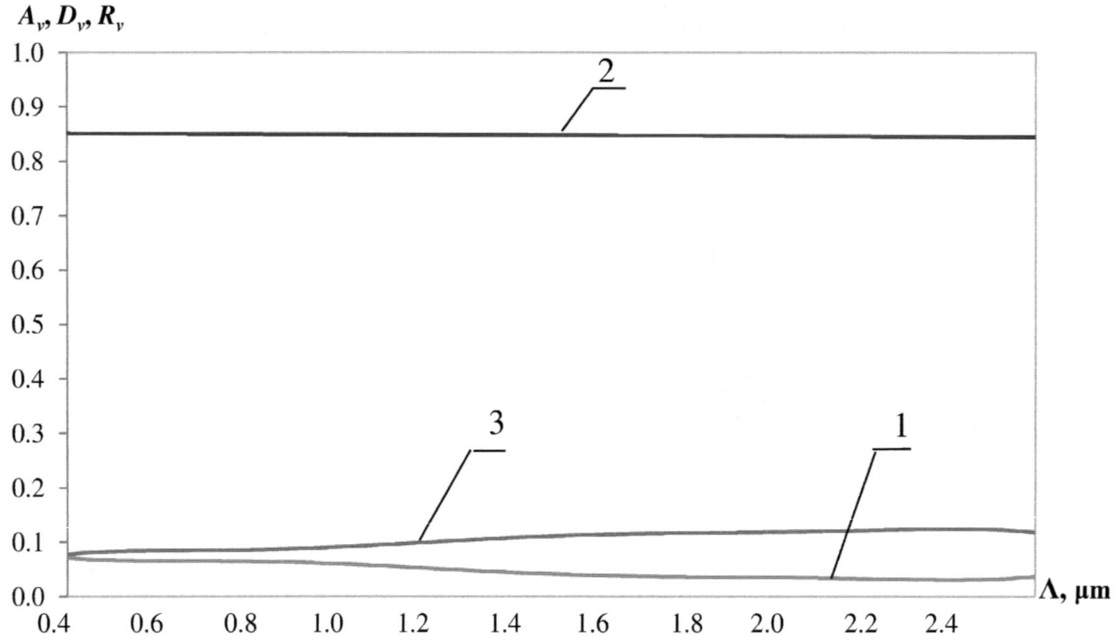

Figure 3. Tungsten mesh spectral optical characteristics
1 – reflectivity R_υ; 2 – transmittivity D_υ; 3 – absorptivity A_υ.

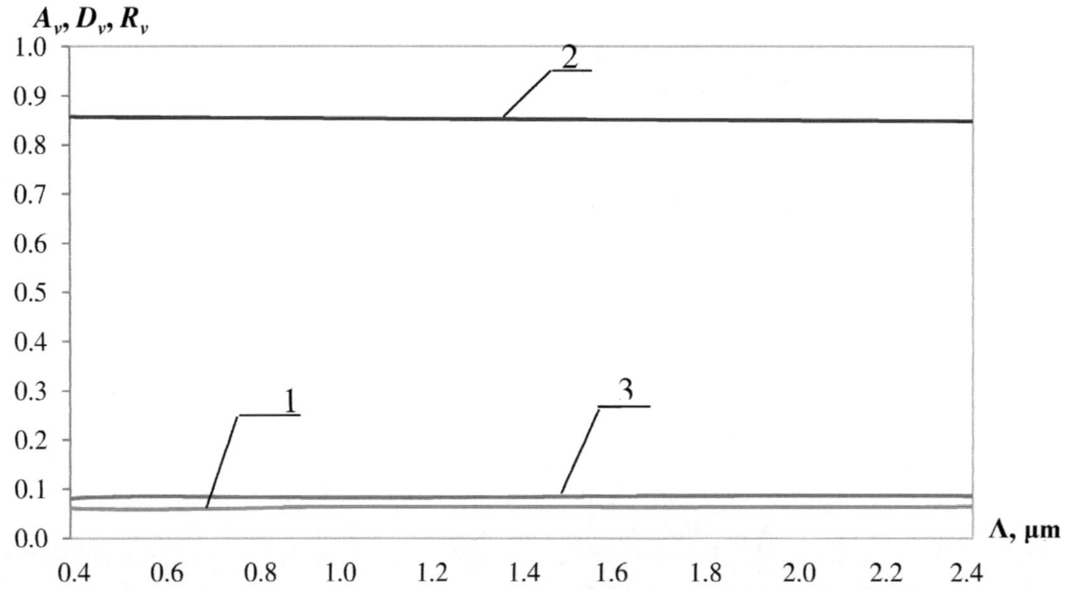

Figure 4. Gold-plated tungsten mesh spectral optical characteristics
1 – reflectivity R_υ; 2 – transmittivity D_υ; 3 – absorptivity A_υ.

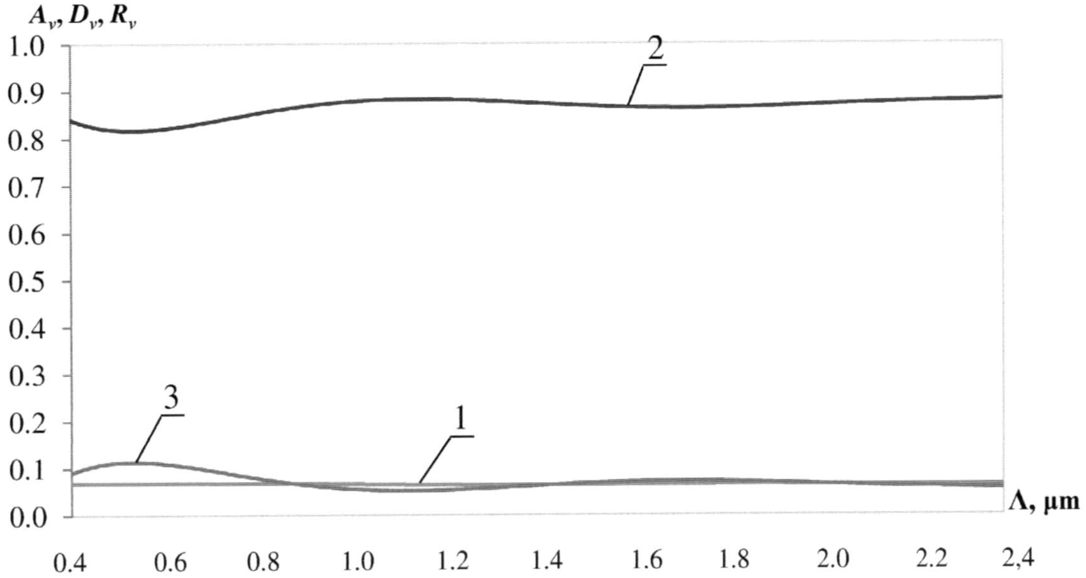

Figure 5. Molybdenum mesh spectral optical characteristics
1 – reflectivity R_υ; 2 – transmittivity $D_\nu D_\upsilon$; 3 – absorptivity A_υ.

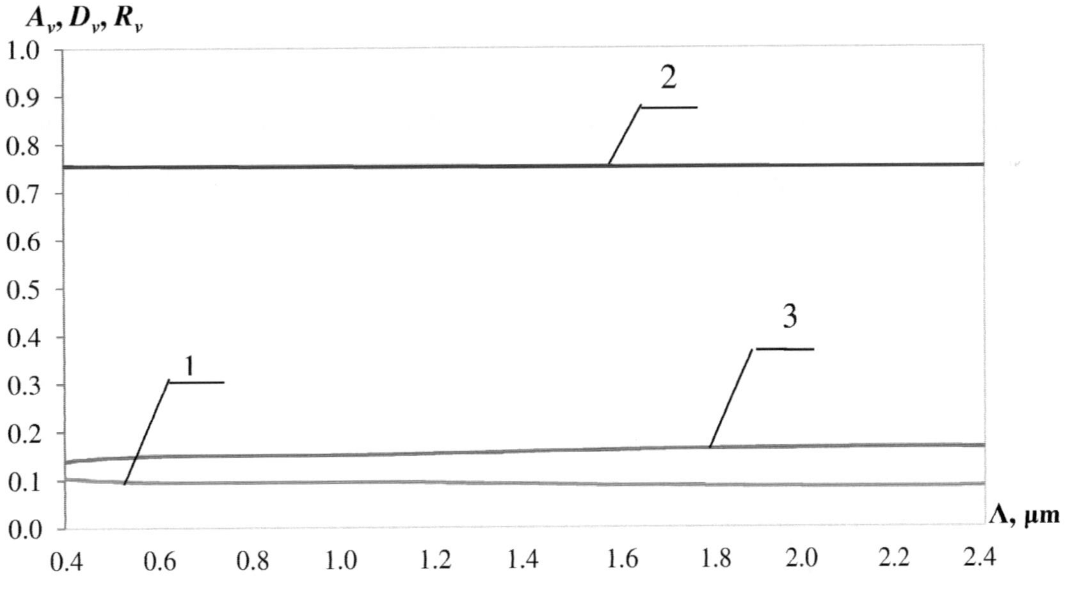

Figure 6. Nickel-coated steel mesh spectral optical characteristics
1 – reflectivity R_υ; 2 – transmittivity D_υ; 3 – absorptivity A_υ.

6. Conclusion
The optical characteristics of four types of metal mesh were experimentally determined: 15-micron tungsten, 15-micron gold-plated tungsten, 20-micron molybdenum, and 50-micron nickel-coated steel wires at normal incidence beam. Based on these data we can draw conclusions about the feasibility of using one of the studied meshes when creating large size deployable reflector. The lowest absorptivity over the greatest part of the spectrum was demonstrated by the gold-plated tungsten mesh, which indicates an advantage for solar radiative heating at GEO.

The reflectivity of all the samples except gold-plated tungsten mesh decreased gradually with the wavelength increasing. The greatest reflectance values were registered for steel mesh (more than 0.8).

Transmittivity of all meshes except molybdenum decreased gradually with the wavelength increasing, but remained in the range 0.75-0.85. The transmittivity of the molybdenum mesh increased slightly. The absorptivity of all meshes remained almost unchanged throughout the range and did not exceed 0.2.

7. Acknowledgment
Some results of this work were obtained in the framework of the Agreement on grant No. 14.577.21.0129 of the Ministry of education and science of the Russian Federation. Unique identifier for applied scientific research (project) RFMEFI57714X0129

References
[1] Archer JS 1980 "High performance parabolic antenna reflectors" *Journal of Spacecraft and Rockets* **17**(1) 20-26. DOI: 10.2514/3.57702
[2] Fager JA 1980 "Large space erectable antenna stiffness requirements" *Journal of Spacecraft and Rockets* **17**(2) 86-92. DOI: 10.2514/3.57712
[3] Thomson MW 2002 *Proc. 20th AIAA International Communications Satellite Systems Conference ICSSC-2002 (Montreal, Canada),* AIAA-2002-2032
[4] Marks G, Lillie C and Kuehn S 2011 Proc. 33rd ESA Antenna Workshop on Challenges for Space Antenna Systems. Preparing for the Future (Noordwijk, The Netherlands)
[5] Semler D, Tulintseff A, Sorrell R and Marshburn J 2010 Proc. 28th AIAA International Communications Satellite Systems Conference -ICSSC-2010 (Anaheim, California, USA), AIAA-2010-8855
[6] Ihle A et al 2012 Proc. Workshop on Large Deployable Antennas (Noordwijk, The Netherlands)
[7] Sinn T, Proetti Zola P, Pfeiffer E K, Datashvily L and Lori M 2015 Proc. 36th ESA Antenna Workshop on Antennas and RF Systems for Space Science (Noordwijk, The Netherlands)
[8] Meguro A, Harada S and Ueba M 2005 Proc. 56th International Astronautical Congress (Fukuoka, Japan), IAF-05-C2.1.B
[9] Harada S, Meguro A and Ueba M 2007 *NTT Technical Review* **5**(1) 70
[10] Ozawa S, Shintate K and Tsujihata A 2012 Proc. Workshop on Large Deployable Antennas (Noordwijk, The Netherlands)
[11] Zimin V N and Sdobnikov A N 2010 "Особенности моделирования динамики крупногабаритных трансформируемых космических конструкций" *Труды XIV Международ. научн. конф., посвящ. памяти ген. конструктора ракетно-космических систем акад. М.Ф. Решетнева* **1** 57-58 ["The features of large transformable space structures dynamic simulation" *Proc. XIV Int. Sci. Conf. dedicated to Acad. M.F. Reshetnev (Krasnoyarsk, Russia)]* **1** 57-58 (in Russian)]
[12] Gottero M, Sacchi E, Lorenzo G, Reznik S V and Kalinin D Yu 2005 *Proc. 35-th Int. Conf. on Environmental Systems (Roma, Italy)*
[13] Evdokimov AS, Ponomarev SV and Buyanov YuI 2011 "Совместный расчет напряженно-деформированного состояния и диаграммы направленности космических рефлекторов"*] Вестник Томского государственного университета. Математика и механика* **1**(13) 74-82 ["The joint calculation of the stress-strain state and the directivity

pattern of the space reflectors" *Herald of Tomsk State University. Mathematic and Mechanics* **1(13)** 74-82 (in Russian)]

[14] Kisanov YA, Feyzulla NM, Kudryavin LA, Zavaruev A 1981 "Материалы для отражательных поверхностей космических складных антенн (КСА)" *Антенны* **29** 20-25 ["Materials for reflectors of folding space antennas (FSA)" *Antennas* **29** 20-25 (in Russian)]

[15] Belyaev OF, *et al* 2007 "Трикотажные металлические сетеполотна для отражающей поверхности трансформируемых наземных и космических антенн" Технический текстиль **16** 59-64 ["Knitted metal meshes to the reflecting surface of the transformable ground and space antennas" *Technical textiles* **16** 59-64 (in Russian)]

[16] Kudryavin LA 1975 Основы теории строения, свойств и процессов выработки сетеизделий трикотажных переплетений Дисс. на соиск. учен. степени д.т.н. Москва: МТИ [*Fundamentals of the theory of structure, properties and processes of the production of knitted* weave *interlacings sitesdaily"* Thesis of Doctor of Science (Technique) Moscow: Moscow Textile Institute (in Russian)]

[17] Zavaruev VA 1980 Исследование особенностей переработки металлических мононитей на вязальных машинах с целью получения полотен технического назначения Дисс. на соиск. учен. степ. к.т.н. [*Research of the features of metallic monofilaments processing on knitting machines for obtaining cloths for technical* applications" Thesis of Candidate of Science (Technique) Moscow: Moscow Textile Institute (in Russian)].

[18] Zavaruev VA 2006 Разработка технологии производства трикотажных металлических сеток для космических и наземных систем связи Дисс. на соиск. учен. степ. д.т.н. [*Development of knitted metal meshes technology for the products of space and terrestrial communication systems"* Thesis of Doctor of Science (Technique) Moscow: Moscow State Textile University n.a. N.A. Kosigin (in Russian)]

[19] Stigene LYa 1990 Дисс. на соиск. учен. степ. к.т.н. Москва: МТИ *The development of structures of knitted fabrics for technical applications from metal threads for the purpose of optimization of their physical-mechanical characteristics"* Thesis of Candidate of Science (Technique) Moscow: Moscow Textile Institute (in Russian)]

[20] Ritikova IV 2005 Разработка технологии формирования сложноконструктивных изделий из металлических трикотажных полотен Дисс. на соиск. учен. степ. к.т.н. [Development of technology of formation of complex structural metal products knitted fabrics for technical purposes" Thesis of Candidate of Science (Technique) Moscow: Moscow State Textile University n.a. N.A. Kosigin (in Russian)]

[21] Kotovich OS 2008 Проектирование структур, свойств и технологии металлических основовязаных сетеполотен для гибких отражательных поверхностей антенн Дисс. на соиск. учен. степ. к.т.н. [The design of the structures, properties and technologies of knitted metal meshes for flexible reflective surfaces of antennas" Thesis of Candidate of Science (Technique) Moscow: Moscow State Textile University n.a. N.A. Kosigin (in Russian)]

[22] Denisova LV, Reznik SV, Kalinin DYu 2011 "Теоретические и экспериментальные исследования тепловых режимов сетчатых рефлекторов космических антенн" Вестник МГТУ им. Н.Э. Баумана. Серия Машиностроение **1**(82) 92-105 ["Theoretical and experimental studies of space antennas metal meshes reflectors" *Herald of Bauman Moscow State Technical university. Machine building* **1**(82) 92-105 (in Russian)]

[23] Reznik SV, Denisova LV, and Churkin VL 1991 "Исследование радиационного теплообмена в металлических сетеполотнах космических антенн" Research report No. 18/91 (Moscow: Joint-Stock Company "Automated Information Systems and Technologies" (AIST) ["Study of radiation heat transfer in metal meshes of space antennas" (in Russian)]

[24] Vasil'ev VN, Mishin GS and Ryabov PM 2004 "Экспериментальное исследование оптических характеристик сетеполотна и элементов антенны КА" *Актуальные*

проблемы российской космонавтики: Труды XXVIII Академич. чтений по космонавтике. М.: Комис. РАН по разраб. научн. наследия пионеров освоения космич. Пространства ["Experimental study of the optical characteristics of metal meshes and elements of the space vehicle antenna" *Proc. XXVIII Academic readings on Cosmonautics* 360 (in Russian)]

[25] 2012 *Fourier transform IR spectrometer Nicolet iS50* (Intertech corporation) http://intertech-corp.ru/aboutproduct.asp?gr=15&subgr=33&prid=231

[26] 2012 *Method of Fourier Transform Infrared Spectroscopy* (ROSNANO) http://rusnano-mc.com/ru/node/229

IOP Conf. Series: Materials Science and Engineering **153** (2016) 012014 doi:10.1088/1757-899X/153/1/012014

Hyperplasticity effect under magnetic pulse straightening of dual phase steel

AP Falaleev[1], VV Meshkov[2] and A Shymchenko[2] Peter the Great St.
[1] Petersburg Polytechnic University, Russian Federation; Sevastopol State
[2] University, Russian Federation.

e-mail: a_falaleev@mail.ru.

Abstract. An investigation of the behaviour of dual phase steel parts during straightening operations, by means of magnetic pulse treatment, is presented. The mechanical analysis of magnetic-pulse treatment for the straightening of thin-walled sheet metal parts produced from dual phase steel was performed, taking into account the effect of hyperplasticity under the influence of the magnetic field. Taking account of the causes of the hyperplasticity and thus the increase of material plasticity, it has been shown that the magnetic impulse gravity can be adjusted by controlling the operation modes. The dependence of the generated magnetic impulse gravity force on the electrical current strength inducted in this part was explored and used for analysis of the magnetic pulse straightening of dual phase steel part. Experimental results were obtained for thin-walled sheet metal part produced from dual phase steel DP 780. The results are used to demonstrate the material deformation under the influence of magnetic impulse gravity force considering the increase of material plasticity. The dependence of relative material deformation on the generated magnetic impulse gravity as well as on the current strength induced in this material was obtained and analyzed.

1. Introduction

Modern automotive industry is faced with various standards and requirements for safety, fuel efficiency and the atmospheric pollution. One possible solution of these problems is the improvement of the automobile body engineering. Such an improvement should take into account not only development and creation of new designs, but also the use of modern materials.

The traditional requirements of the vehicle body are high strength, in both static and dynamic terms; and impact stability of body shape, and stiffness, which is defined by elastic modulus of material. For a long time mild steel has met all these requirements and has been the main and basic material of the automobile industry. The main reason for this that mild steel has advantageous properties such as drawability and work-hardening; parameters that characterize the material behaviour during production processes such as forming and stretching.

These better properties, as well as lower cost and well-developed production technologies, ensured the leader place for mild steel [1]; however, modern developments in metallurgy and materials science have lead to the availability of a wide range of materials offering different sets of mechanical and physical properties. In some circumstances, the new plastics and composite materials can provide better properties than traditional steels, and, as a result of lower density, the usage of these materials in automobile construction can also enable a vehicle weight reduction. In turn, this can result in improved indices of fuel efficiency and atmospheric pollution. Various materials such as aluminium, magnesium and titanium also offer potential weight reduction and improvement of mechanical properties of automobile body. Compared with plastics and composites, light alloys are much more preferable for application in mass-production vehicles due to the significant development of processing and manufacturing capabilities for products made from these alloys.

Despite the wide range of materials, the main choice of the automobile manufacturers is still steels. This preference can be explained by issues of production cost, passive safety, mass-production, and the simplicity of repair and restoration. The weight reduction in this case is achieved by the improvement of design and usage of Advanced High Strength Steels (AHSS), such as Complex Phase steels, Boron steels, Transformation Induced Plasticity (TRIP) and Dual Phase (DP) steels. Although these steels do not show a weight reduction relative to traditional mild steel, the high value of strength of these materials allows a *pro rata* reduction in thickness of automobile body elements that then leads to the weight reduction. The usage of AHSS enables the continued utilization of the existing production and manufacturing technologies [1].

Among the types of AHSS, the dual phase steel should be highlighted as the steel having the best formability and elongation compared with other high-strength steels of the same strength. The microstructure of dual-phase steels commonly consists of a ductile phase matrix (ferrite) with some amount (about 15%) of hard martensite phase induced by heat treatment and alloying processes. Control of the martensite concentration enables the strength of the material to be determined at any value from 350 to 1000 MPa, while the presence of ferrite in the microstructure is responsible for material ductility [2, 3]. As a result, the dual-phase steel combines both high strength and high relative elongation properties, resulting in high energy absorption, as well as good formability and the possibility of processing by the cold forming method.

As described, it is no surprise that a wide range of grades of dual-phase steel is used for automotive industry such as DP500, DP600, DP780 and DP980; however, the parts made from dual phase steel or AHSS have some limitations and problems in the stages of production and manufacturing. These problems are related to the dependence of the strength and the plastic behaviour of these materials on the strain and heat treatment history. The technological memory leads to significant hardening of the material and thus to complications during restoration or forming. For example, Hansen [4] showed that a relatively small increase of plastic strain from 3 to 5 % leads to an increase of the yield strength of dual-phase steel from 70 to 80 ksi (from 482 to 552 MPa). This hardening presents a need to increase manufacturing operation pressure forces and significantly increase operation time during the straightening process of automobile body, after damage or traffic accidents. The presence of some residual strain is also related to the pre-strain effect incurred during the manufacturing stages. It is well known that the amount of pre-strain, and as result, the plastic behaviour, depends on the loading path [5, 6]. Most parts of automobile body elements are fabricated using sheet metal forming processes that very often include changes in the loading path. For example, the material flowing through the die corner moving to die, bending and unbending in different directions, complex deformation due to the complicated configuration of die and other impact during multi-stage fabrication processes such as deep-drawing and blanking. The result the need for complex loading paths during manufacture processing leads directly to the presence of some pre-strain [7, 2].

Huh *et al.* [2] show that the some amount of pre-strain effects the dynamic tensile properties, the initial yield stress and the flow stress. According to the presented results the yield strength and tensile strength of DP600 steel, with pre-strain of about 10%, increased by 22.6 MPa and 32.8 MPa respectively at the strain rate of 100 s^{-1}.

Morestin and Boivin [8] describe a decrease in elastic modulus during the unloading of dual-phase sheets, leading to non-linear strain recovery, which has an undesirable effect during forming processes. According to their results, an increase in plastic strain by 5% leads to a decrease of 10% in the unloading elastic modulus; moreover, the unloading modulus will degrade with increasing plastic pre-strain [9, 10].

Technology such as Superplastic Forming (SPF) provides an alternative approach to material forming. The superplasticity or hyperplasticity is characterized by elevated ductility, with such materials displaying elongations of up to 200% (for some materials the elongation during SPF can reach the value 400% and higher). The benefits of SPF usage are a reduction in residual stresses and in springback. The mandatory conditions for SPF process conditions are [11]:

- The presence of two or more fractions in the microstructure
- The presence of a fine stable grain size, usually between 5 – 15 μm
- Deformation being made at a relatively slow strain rate, normally in the range from 10^{-4} s^{-1} to 10^{-2} s^{-1}
- An operation temperature higher or equal to 0.5 T_m, where T_m is the melting temperature of processed material, measured in Kelvin

On the other hand, SPF is a slow process compared with the traditional processes of metal sheet forming [11].

Unfortunately, not all materials are suitable for this type of forming. Commonly, the SPF process is used for processing of various aluminium, magnesium, and titanium alloys [11-14]; however, in the work [15] it has been established that ultrafine grained ferrite/martensite dual phase steel, produced by an equal channel angular pressing (ECAP) process, can also be processed as a superplastic material.

Another way to obtain the superplasticity during the forming process is to use a magnetic field, and an increase in material plasticity and a reduction in residual stresses has been demonstrated [16, 17]. It has been shown, [18], that the elongation of HS 37/23 steel can be increased by between 4.95% and 7.50% under the influence of a magnetic field with magnetic inductions of 0.4 T and 1.07 T respectively. Golovashchenko *et al.* [19] demonstrated a method for elimination of springback by applying an electromagnetic force on a workpiece that is fixed in the die. According to their results, about 90% of internal stresses can be eliminated by applying this method. These results were confirmed by Iriondo *et al.* [20] by testing the sheet samples made from DP600 and TRIP700 high strength steels and applying electromagnetic discharges of from 8 to 24 kJ. The changes in observed sizes of U-channels samples was about 8 mm compared with samples formed without applying an electromagnetic field.

Among the technologies of magnetic forming of metal sheets, there is the method that can be used for straightening of automobile body [21, 22]. This method is based on Ampere's law and the interaction between two conductors with the current flow, while placed in a magnetic field. The main purpose of this article is to study the behaviour of dual-phase DP780 steel during this forming process, and to examine the presence of the superplasticity/hyperplasticity.

2. Experimental method

2.1. Materials parameters

The material investigated in this study was dual phase DP780 Y450 steel sheet, from which samples were taken. The experimental measurements were carried out three times. The thickness of each sample tested was 0.001 m. The chemical composition and mechanical properties of each dual phase steel studied are shown in Tables 1 and 2. The microstructure of the material comprises ferrite (about 85%) and martensite fractions. The samples were tested without pre-strain or heat treatment preparations.

2.2 Experimental parameters

Figure 1 represents schematically the experimental setup used for this particular study. Eddy currents are induced in the sheet metal blank, as well as in the accessory attracting screen, by the discharge of an electromagnetic impulse generated by the coil inductor [22]. The induction of current in the conductive material placed in the variable electromagnetic field leads to the appearance of attracting and repelling forces. It has been established [21] that the frequency of electromagnetic field has significant influence on the summation action of these two forces. According to the results obtained, the repelling force is slight and can be neglected at the frequency ≤ 2 kHz.

For the study of the behaviour of dual-phase steel, for applications such as repair and maintenance of sheet metal constructions (straightening process, dent removing, etc.), the test frequency chosen was 2 kHz. The single turn induction coil radii were $R_1 = 0.07$ m and $R_2 = 0.075$ m. The experiment was carried out with discharge times of 0.1, 2.5 and 5 μs, and maximum current I_{max} = 70 to 82 kA. The values of the current induced on the surface of the sheet metal blank and on the accessory attracting screen are measured using an Ammeter. The attractive force acting on the sheet metal blank can be calculated by equation (1) [21, 22]:

$$F_{\text{attr}} = \mu_0 \cdot J_{\text{ind}}^b \cdot J_{\text{ind}}^s \cdot \frac{r}{(2h)}, \text{N} \quad (1)$$

where J_{ind}^b, J_{ind}^s (A) is the current induced in the sheet metal blank and in the accessory attracting screen, respectively; r is the radius of the operation area (in the range from 0 to R_1 = 0.07 m); and h is the distance between the coil inductor, the accessory attracting screen and the sheet metal blank. In this particular study, the distance h is 0.15 mm. $\mu_0 = 1.26 \times 10^{-6}$ N A^{-2} is the permeability of free space.

Table 1. Chemical composition of the dual-phase DP780 Y450 steel (wt.%).

Steel grade	C	Mn	P	S	Si	Cr	Mo	Cu	W
DP780 Y450	0.17	2.2	0.007	0.6	0.010	0.399	0.270	0.001	0.005

Table 2. Mechanical properties of dual-phase DP780 steel.

Steel grade	Tensile strength (MPa)	Yield strength (MPa)	Strain rate (s^{-1})	Elongation (%)
DP780 Y450	842	450	0.63	18.5

Table 3. The results of experimental measurements.

J_{ind}^b (kA)	t = 0.1 (μs)			t = 2.5 (μs)			t = 5 (μs)		
	ε1, %	ε2, %	ε3, %	ε1, %	ε2, %	ε3, %	ε1, %	ε2, %	ε3, %
70	0.50	0.54	0.48	0.58	0.63	0.60	0.71	0.80	0.68
72	0.54	0.56	0.53	0.66	0.67	0.64	0.79	0.77	0.76
74	0.61	0.61	0.59	0.74	0.75	0.74	0.85	0.85	0.83
76	0.72	0.70	0.75	0.89	0.84	0.77	0.97	1.04	0.92
78	0.98	0.96	0.98	1.05	1.10	1.07	1.20	1.15	1.17
80	1.21	1.25	1.22	1.30	1.30	1.27	1.44	1.42	1.41
82	1.53	1.52	1.50	1.62	1.70	1.58	1.70	1.72	1.72

The tensometric linear displacement sensor is used to define and measure the strain related to appropriate current induced in the system being studied. This sensor has a sensitivity of about 0.001 mm, which enables precise control of the material deformations.

The temperature measurements during magnetic pulse straightening are performed by means of a laser pyrometer, "MLG 36.6 Double". The range of its measurement is from -50 to +250 ^{0}C and the measurement error is less than 1.5%.

The hardness of the tested samples was measured using the Vickers hardness test, as is consistent with the ISO 6507-1 standard.

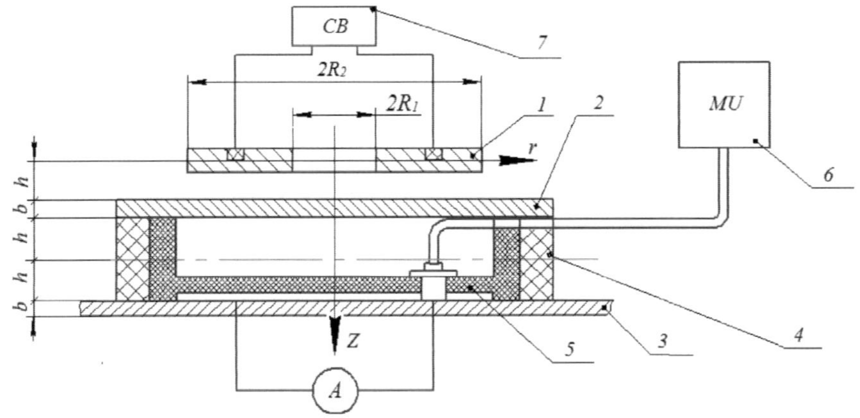

Figure 1. Schematic of experimental setup:
1 – Multi-turn induction coil; 2 – Accessory attracting screen; 3 – Sheet metal blank;
4 – Dielectric intermediate piece; 5 – Spacer with the tensometric linear displacement sensor;
6 – Measuring unit; 7 – Control block.

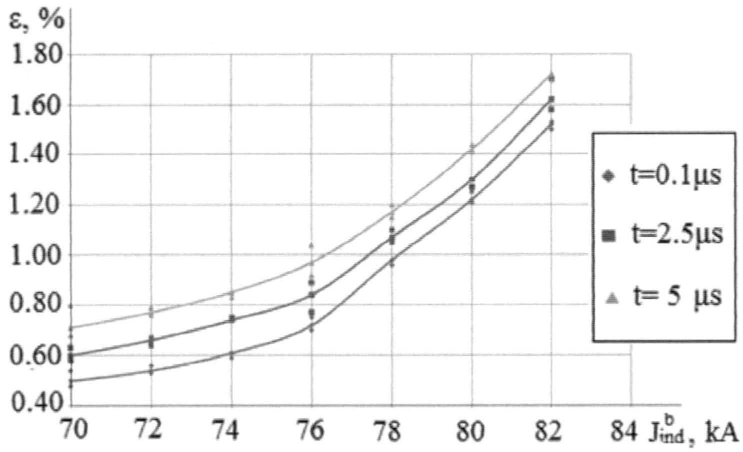

Figure 2. The strain – current relationship curves for t = 0.1, 2.5 and 5 μs.

3. Result and Discussion

During the experimental measurements the relations of relative deformation from the induced current strength in sheet metal blank were obtained. The strain results, for each of the three tests for the three pulse durations, are presented in Table 3.

The experiment was carried out three times for each value of the discharge time (t) in order to take into account any inaccuracy in measurements. The relation between strain and current for each value of the discharge time is presented in figure 2.

It is also worth noting that the value of induced current in the sheet metal blank and in the accessory attracting screen is non-uniform through the operation area. The current distribution through the operation area is presented in the figure 3.

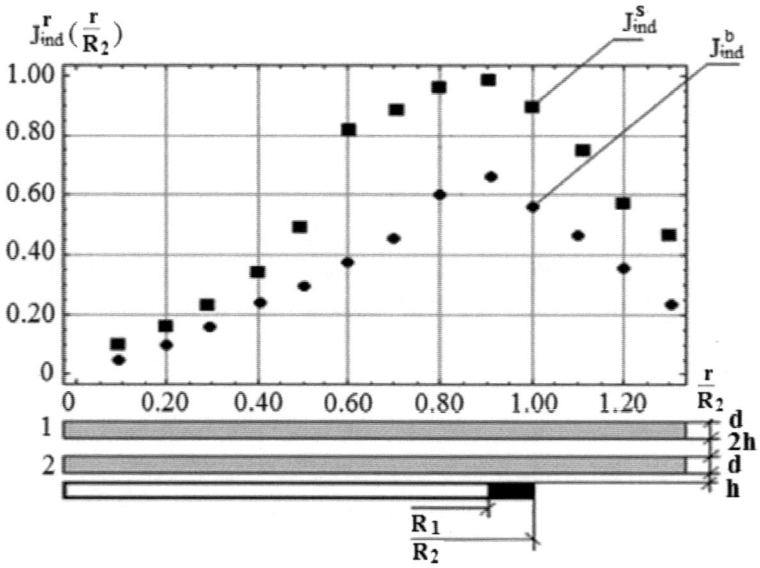

Figure 3. The current distribution through the operation area.

The results show the maximum obtained strain for the maximum current. According to figure 3, the current induced in the blank increases with distance from the centre of the operating area, and reaches a maximum when the distance r is equal to the R_1 parameter. In order to simplify the further calculation, the measured strain is assigned to the narrow section from $r_1 = 0.8R_2$ to $r_2 = 0.93R_2$. The attractive force on this area is the average value between force at the distances r_1 and r_2, which can be calculated by using equation (1).

The stress applied on this area by attractive force is given by:

$$\sigma = \frac{F_{attr}}{A}, \text{ MPa} \tag{2}$$

where F_{attr} (N) is the average attractive force between tested sample and the accessory attracting screen, and A is the surface area of the sample part, which is exposed to the influence of the attractive force F_{attr}. Considering the simplification made earlier, the surface area is:

$$A = \pi \cdot R_2(0.93 - 0.8), \text{ m}^2 \tag{3}$$

The results of calculations are presented in Table 4 and the stress – strain curve is presented in figure 4.

The stress-strain curves obtained represent the plastic deformation region. The plastic deformation is more important for straightening operations and dent removal, therefore the elastic region was neglected during these experimental measurements. From the results it can be seen that the strength of the samples straightened with a pulse with a discharge time of about 5 μs is lower compared with that of the other samples. The ductility did not change significantly.

The probable reason of this effect can be ascribed to the influence of the magnetic field on the material microstructure, leading to a change in the amount and distribution of the microstructural defects that take part in work hardening process. During plastic deformation of crystalline material the microstructural defects, such as vacancy defects, itinerant electrons, *etc.*, deform the crystal lattice of the material and prevent the dislocation movements that lead to material hardening, and the increase of internal stresses. Under the influence of the magnetic field, the number of internal defects decreases as a result of their self-elimination under the action of the Lorentz force. These changes lead to a reduction in the barriers to dislocation movement and thereby increase the material plasticity.

Table 4. Stress-strain relation for tested material.

F_{attr} (kN)	σ (MPa)	t = 0.1 (μs)			t = 2.5 (μs)			t = 5 (μs)		
		ε1 (%)	ε2 (%)	ε3 (%)	ε1 (%)	ε2 (%)	ε3 (%)	ε1 (%)	ε2 (%)	ε3 (%)
1926.3	481.57	0.50	0.54	0.48	0.58	0.63	0.60	0.71	0.80	0.68
2037.9	509.48	0.54	0.56	0.53	0.66	0.67	0.64	0.79	0.77	0.76
2152.7	538.18	0.61	0.61	0.59	0.74	0.75	0.74	0.85	0.85	0.83
2270.7	567.67	0.72	0.70	0.75	0.89	0.84	0.77	0.97	1.04	0.92
2391.7	597.94	0.98	0.96	0.98	1.05	1.10	1.07	1.20	1.15	1.17
2515.9	629.00	1.21	1.25	1.22	1.30	1.30	1.27	1.44	1.42	1.41
2643.3	660.84	1.53	1.52	1.50	1.62	1.70	1.58	1.70	1.72	1.72

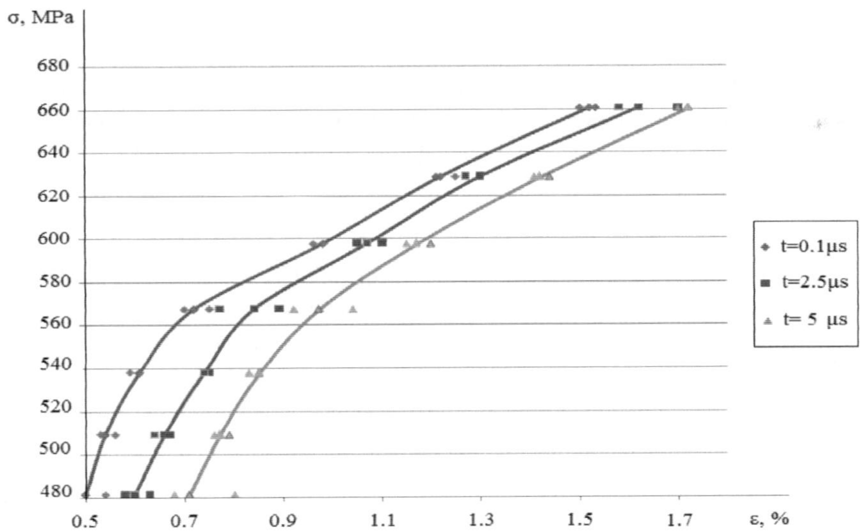

Figure 4. Stress-strain curves of DP780 steel under pulsed action during discharge times 0.1, 2.5 and 5 μs.

Note that there is some concave area on the stress-strain curves for discharge times of about 0.1 μs and 2.5 μs. For a discharge time of 5 μs this area is poorly expressed. This concave area is characterized by a decrease in the material strength and an increase in elongation. The non-typical part of curve detects the improvement in sample plasticity, in the loading range between about 561 and 600 MPa.

Knowing the relationship between the values of induced current and applied loading (equations 1 and 2), it is possible to find the current range responsible for the change in material properties. Calculations showed that, for a current density of about 1071 $A \cdot mm^{-2}$, the onset of the concave area is observed. This result is in agreement with previous works [17, 23], where the authors noted that the increase of material plasticity is more noticeable for a pulse duration of about 10^{-4} s and current density of between about 10^3 to 10^4 $A \cdot mm^{-2}$; however, this variation of curve shape might also indicate material hardening and accumulation of some internal stresses.

In order to find the reason for these results, the hardness of DP780 steel samples was measured. The hardness not only demonstrates the properties of material surface, but also provides an evaluation of the ability of material to resist plastic deformation. Measurements of sample hardness by the Vickers method gave results within the range 245 to 250 Hv, while the typical hardness of DP780 is about 250 Hv. The hardness of samples straightened under a discharge time of $t = 0.1$ μs and current density of about 1071 $A \cdot mm^{-2}$ does not exceed the normal value of DP780 steel hardness, and is 246 Hv. Based on the results obtained, it can be assumed that there is no hardening process or stress accumulation, thus, the concave area indicates the presence of superplasticity under the influence of pulsed magnetic field.

One of the main purposes of this article is to define the parameters that have an influence on the results of the magnetic pulse straightening and can be used as control and operation parameters. It is well known that temperature changes lead to vibrations of crystal lattice, thus the operation of temperature is one of the possible parameters. Measurement of the temperature of samples during the experiment was attempted; however, the thermal inertia of the dual-phase DP780 steel did not allow for effective direct temperature measurement. It must be noted that extra heating was not used during the experiment: any material heating is restricted only to that generated locally by the flow of the induced current through the conductive material. Since the range of induced current for each test was the same (from 70 to 82 kA), the temperature conditions were similar for all test measurements; however, the values of strain obtained differ for each duration of the discharge time. Based on this, it is possible to assume that the operation of temperature has less of an effect on the dual-phase material deformation compared with the other control parameters under magnetic pulse action. Additional research is required in order to evaluate the role of the operation of temperature during this kind of material forming.

It was observed that the discharge time has some influence on the material strength and deformation. The discharge time is responsible for both the strain rate and the duration of magnetic field treatment. Since the strain rate did not show a significant effect on the material deformation, the variations in results can be attributed to the difference in the duration of the magnetic field influence. It can be assumed that the increase of the discharge time and magnetic field impact can lead to weakening and relaxation of DP780 steel material. This assumption agrees with the results obtained in works [19, 20, 24].

4. Conclusion

Magnetic pulse technology is a potentially available method for straightening and repair of automobile body elements. The results obtained show the ability of the proposed system to apply magnetic force and form the sheet metal elements; moreover, the application of magnetic field during the straightening process has some influence on the structure of dual-phase steel materials. This influence provides the necessary decrease in material strength during the forming process to circumvent the effects related with internal stresses, pre-strain history, and work hardening during loading – unloading cycles. This effect is similar to the effect of hyperplasticity, which is observed during Superplastic Forming (SPF) processes as well as during forming operations employing a magnetic field.

The relationship between relative deformation, induced current and magnetic force were obtained. According to these results, the induced current discharge time had the more significant influence on the material deformation. This parameter, as well as the strength of induced current, can be taken as the main operation parameters for the control of the straightening process. It was shown that the improvement of DP780 steel plasticity is more significant for pulse durations of between about 0.1 and 2.5 μs, and with a current density from 1071 A·mm^{-2}.

The other parameters, such as the strain rate and the operation temperature, did not show a significant effect; however, their real role for the magnetic pulse straightening of sheet metal elements should be evaluated on the basis of further researches.

5. Acknowledgments

This research was funded by the Ministry of Education and Science of the Russian Federation according to the base part of state tasks of the Sevastopol State University №2014/702 (Project 3867).

References

[1] Davies G 2012 "*Materials for Automobile Bodies*" (Oxford OX5 1GB, UK: Elsevier)
[2] Huh H, Kim SB, Song JH and Lim JH 2008 "Dynamic tensile characteristics of TRIP-type and DP-type steel sheets for an auto-body" *International Journal of Mechanical Sciences* **50** 918–31
[3] Ishikawa N *et al.* 2015 "Microscopic deformation and strain hardening analysis of ferrite-bainite dual-phase steels using micro-grid method" *Acta Materialia* **97** 257–68
[4] Hansen SS 1982 "The Formability of Dual-Phase Steels" *Journal of Applied Metalworking* **2**(2) 107–18
[5] Tutyshkin N, Muller WH and Zapara M 2014 "Strain-induced damage of metals under large plastic deformation: Theoretical framework and experiments" *International Journal of Plasticity* **59** 133–51
[6] Barlat F *et al.* 2014 "Enchancements of homogenous anisotropic hardening model and application to mild and dual-phase steels" *International Journal of Plasticity* **58** 201–18
[7] Yu HY and Shen JY 2014 "Evolution of mechanical properties for a dual-phase steel subjected to different loading paths" *Materials and Design* **63** 412–8
[8] Morestin F and Boivin M 1996 "On the necessity of taking into account the variation in the Young modulus with plastic strain in elastic-plastic software" *Nuclear Engineering and Design* **162**(1) 107–16
[9] Govik A, Rentmeester R and Nilsson L 2014 "A study of the unloading behaviour of dual phase steel" *Materials Science & Engineering A* **602** 119–26
[10] Ghaei A, Green DE and Aryanpour A 2015 "Springback simulation of advanced high strength steels considering nonlinear elastic unloading-reloading behavior" *Materials and Design* **88** 461–70
[11] Ridley N 2011 "*Superplastic forming of advanced metallic materials. Methods and applications.*" (Cornwall, UK: Woodhead Publishing Limited) 1–31
[12] Wang CW, Zhao T, Wang G, Gao J and Fang H 2015 "Superplastic forming and diffusion bonding of Ti-22Al-24Nb alloy" *Journal of Materials Processing Technology* **222** 122–127
[13] Guo ML, Liu J, Tan MJ and Chua BW 2014 "Microstructure evolution of Ti-6Al-4V during superplastic-like forming" *Procedia Engineering* **81** 1090–95
[14] Kumaresan G and Kalaichelvan K 2014 "Experimental studies of a rectangular cup formation of Al7075 Alloy in Superplastic Forming Process" *Procedia Material Science* **6** 892–896
[15] Okayasu M, Sato M, Mizuno M, Hwang DY and Shin DH 2008 "Fatigue properties of ultra-fine grained dual phase ferrite/martensite low carbon steel" *International Journal of Fatigue* **30** 1358–65
[16] Малюшевский ПП 1983 *"Основы разрядно-импульсной технологии"* (Киев, СССР: Наукова Думка) [Malyushevsky P P 1983 *"Basics of the pulse technology"* (Kiev, USSR:

Naukova Dumka) (in Russian)]

[17] Делюсто ЛГ 2005 *"Основы прокатки металлов в постоянных магнитных полях"* (Москва, Россия: Машиностроение) [Delyusto L G 2005 *"Fundamentals of metal rolling in a constant magnetic fields"* (Moscow, Russia: Mashinostroyeniye) (in Russian)]

[18] Гринкевич ВА, Шевченко ТН, Краев МВ, Краева ВС, Бондарев СВ 2013 "Экспериментальное исследование пластической деформации стали Ст3 во внешнем магнитном поле" *Обработка металлов давлением* **37**(4) 79–82 [Grinkevich V A, Shevchenko T N, Krayev M V, Krayeva V S and Bondarev S V 2013 "Experimental Study of Plastic Deformation of St3 steel in a magnetic field" *Obrabotka metallov davleniyem* **37**(4) 79–82 (in Russian)]

[19] Golovashchenko S 2005 "Springback calibration using pulsed electromagnetic field" *AIP Conf. Proc.* **778** 284–5

[20] Iriondo E, Alcaraz JL, Daehn GS, Gutierrez MA and Jimbert P 2013 "Shape calibration of high strength metal sheets by electromagnetic forming" *Journal of Manufacturing Processes* **15** 183–93

[21] Batygin YV, Golovashchenko SF, and Gnatov AV 2013 "Pulsed electromagnetic attraction of sheet metals – Fundamentals and perspective applications" *Journal of Materials Processing Technology* **213** 444–52

[22] Batygin YV, Golovashchenko SF and Gnatov AV 2014 "Pulsed electromagnetic attraction of nonmagnetic sheet metals" *Journal of Materials Processing Technology* **214** 390–401

[23] Троицкий ОА 2009 "Ультразвуковое электропластическое плющение металла" *Вестник научно-технического развития* **10**(26) 42–9 [Troitskiy O A 2009 "Ultrasonic electroplastic flattening of metals" *Vestnik nauchno-tekhnicheskogo razvitiya* **10**(26) 42–9 (in Russian)]

[24] Falaleev AP, Meshkov VV, Vetrogon AA and Shymchenko AV 2016 "Evolution of mechanical properties of boron/manganese 22MnB5 steel under magnetic pulse influences" *IOP Conf. Ser.: Mater. Sci. Eng.* **110** 012107

IOP Conf. Series: Materials Science and Engineering **153** (2016) 012015 doi:10.1088/1757-899X/153/1/012015

Theory and practice for the manufacture of a composite thermal heat shield for a space ship

V A Tarasov[1], M A Komkov[1], V A Romanenkov[2], A I Alyamovsky[2], N I Kopyl[2] and R V Boyarskaya[1]

[1] BMSTU, 5, 2nd Baumanskaya Str., Moscow, Russia.
[2] RSCE, 4-A, Lenin Str., Korolev, Moscow Region, Russia.

tarasov_va@mail.ru

Abstract. The technological processes were explored for the manufacture in an autoclave of a space ship heat shield. A mathematical model was created for the determination of the duration of the impregnation of the binder for the composite material. The change in the Nitrogen content is dependent on the time in the autoclave. This dependence relates to the use of the minimum amount of electricity to reduce the expense of the process in practice.

1. Introduction

The development of manned space flight is impossible without solving problems of thermal protection of spaceships and crew during their return to Earth. When entering the dense atmosphere at a rate of 8-12 km/s, the pressure and the temperature on the surface of the descending spaceship (DS) from the side of forward flow reaches high values in the process of adiabatic compression. The duration of landing and the rise of heat flow are sufficient for the destruction of any of the known materials, therefore, in DS a special windshield with ablative thermal protection is used [1-6].

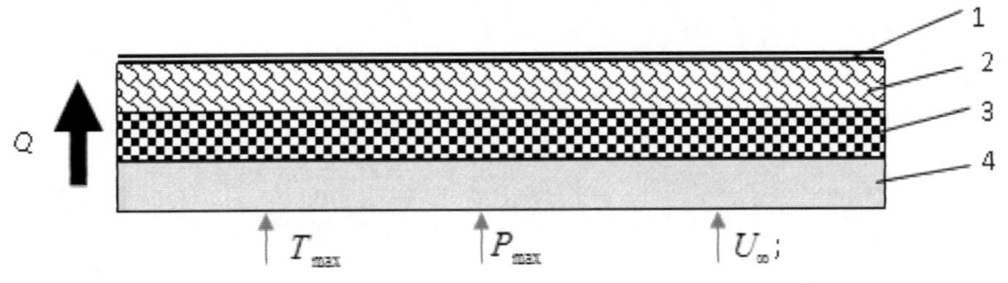

Figure 1. The layout of heat strata within the material of the thermal protection: 1 – DS metal wall; 2 – basic material layer of thermal-protective coating (TPC); 3 – pyrolysis binder layer (separation of liquid and gas phase) and formation of charred residue; 4 –"sacrificial" erosion layer; Q – heat flow, U_∞ – spaceship speed, T_{max}, P_{max} – maximum temperature and pressure occuring during adiabatic compression of air flow at the stagnation point.

The effect of ablative thermal protection is schematically represented in figure 1. Intense heat flow Q causes the formation of several areas in heat-protective material. In area 2, in the direction from design wall 1, the DS material serves as a thermal shield, thus, it must have low thermal conductivity (coefficient of thermal conductivity ≤0.2 W/m K).

The intense power loads have an influence on the material of thermal protection. The material with minimum porosity and compositional structure, including filler and matrix (binder after polymerization) is used. The fibrous filler is a package of woven workpieces sewn in thickness on the base of asbestos, basalt, quartz or kaolin fibers.

In area 3, at high temperature the process of matrix high-molecular polymer decomposition occur with the formation of volatile decomposition products and coke residue. The part of incoming heat is consumed for an endothermic reaction process of polymer transformation. Coke performs a function of an absolute black body in the process of radial heat flux in accordance with Stefan-Boltzmann law. Matrixes of ablative thermal protection type on the basis of phenol-formaldehyde and silicone thermosetting binder are the most effective in relation to the process of coke formation (coke output is up to 65%).

In area 4, at temperature of more than 2000-2500 K the gaseous, liquid and solid phases, which are carried away by the atmospheric incoming air flow, are separated from glowing coke that maintains strength. It is a so-called "sacrificial" layer of DS thermal protection, which is exposed to erosive carry-over. The requirements for the reduction of the thermal protection cover are thickness $U_{\text{carry over}}$ ≤ 0.1 mm/s.

This paper is dedicated to a discussion of the technological features for the provision of DS thermal protection requirements and the prospects for further improvement of the technological processes.

2. Manufacture of a thermal protective coating for a descending spaceship

The production technology of a thermal protection cover for reusable spaceships was developed by the Rocket-Space Corporation "Energy" (RSCE) named after S P Korolyov, which is currently the only organization in Russia that holds and develops the technologies for spaceships and man-controlled space flight.

a b

Figure 2. DS thermal protective shell (a) and frontal heat shield (b).

The development of thermal protection (figure 2) ensured the high aerodynamic quality and reliability of the descending spaceship "Soyuz". Invaluable experience was gained from the thermal protection of the descent vehicle "Zond" thermal protection, designed for manned flights to the Moon and overcoming the Earth's atmosphere during the return with the velocity 11km/s. The results of experimental and theoretical studies were summarized in joint works of RSCE and Bauman Moscow State Technical University (BMSTU) specialists [7-23], including the application of an autoclave to cure and form the composite under the action of heated nitrogen pressure.

The required porosity of the TPC is achieved by means of repeated filling of a porous workpiece with binder and drying. Repeated filling is needed because of the secondary porosity which is formed by the evaporation of alcohol in the binder.

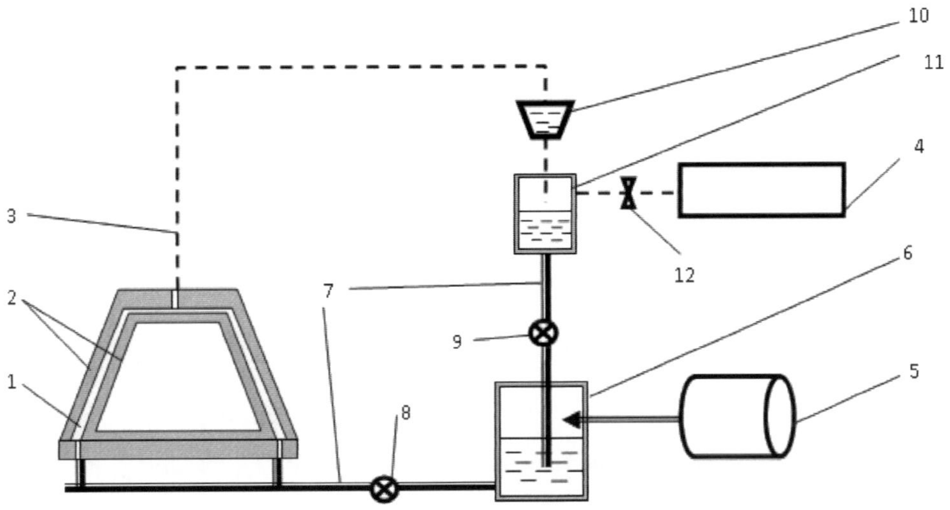

Figure 3. Schematic diagram of apparatus for DS thermal-protective coating impregnating under pressure:
1 – workpiece, 2 – manufacturing jig, 3 – vacuum line, 4 – vacuum pump, 5 – compressed gas bottle, 6 – the binder container, 7 – the pipeline for supplying the binder, 8 and 9 – valves, 10 – binder indicator, 11 – binder collector, 12 – valve.

The technological scheme of filling of a semi-finished product with the binder is shown in figure 3. The workpiece, 1, to be impregnated is placed in the cavity of manufacturing jig, 2, which is connected by a vacuum line, 3, with a vacuum pump, 4. As a result, vacuum in the pore space of the workpiece is produced. Following this, the compressed gas in the bottle, 5, creates in container, 6, with binder the overpressure of displacement. The binder is supplied over pipeline, 7, into pore space of the workpiece when valve, 8, is opened. In order to prevent movement of the binder in the excess binder collector, 11, the valve, 9, is closed. The impregnation process lasts until indicator, 10, shows a complete filling of the pore space with binder. Then, the vacuum pump stops working, valve, 12, is opened ensuring air drainage to the vacuum created in the special capacity of the plant. Then valve, 8, is closed and valve, 9, is opened. The bleed-out from collector 11 drains into binder container 6.

The second operation of the repeated process chain, which is carried out in an electrovacuum chamber furnace or autoclave, is a thermal vacuum drying at a temperature of $T_{dry} = 50 - 60$ °C and pressure $P_{dry} = 10 - 20$ kPa during $\tau_{dry} = 8 - 10$ hours.

After the final impregnation, the semi-finished product is fixed on the mandrel and the thermal treatment lasts 2 – 4 hours at temperature 135 °C. The pressure is set equal to 0.2 MPa when processing a sidewall and 1 MPa when processing a front board. Taking into account the time of temperature rise and its reduction to normal value, the duration of thermal treatment operation takes 18 hours.

The duration of impregnation is important for the organization of technological process. The duration of liquid impregnation of a solid medium with an open porosity is defined in accordance with the Darcy equation [25]:

$$q_p = \frac{K_D}{\eta_C}\frac{\Delta p}{l} \tag{1}$$

where, q_p is the liquid volume proceeding in unit of time through unit of area of section of porous material, η_C is the viscosity of the liquid binder, Δp is the pressure difference inside the solid medium filled with fluid, l is the maximum path length through which the fluid passes when filling the pore space, and K_D is the Darcy-Kozeny-Pocket factor [17].

Alcohol addition to the binder reduces the viscosity and the impregnation process duration D_i (figure 4), but increases the residual porosity P_0 during drying. Initial data for calculation is selected so that the binder content in the TPC finished material would be at 28 – 30%, and the fibre content in the TPC finished material 67 – 70%. The workpiece size intended for impregnation is l = 800 mm.

a

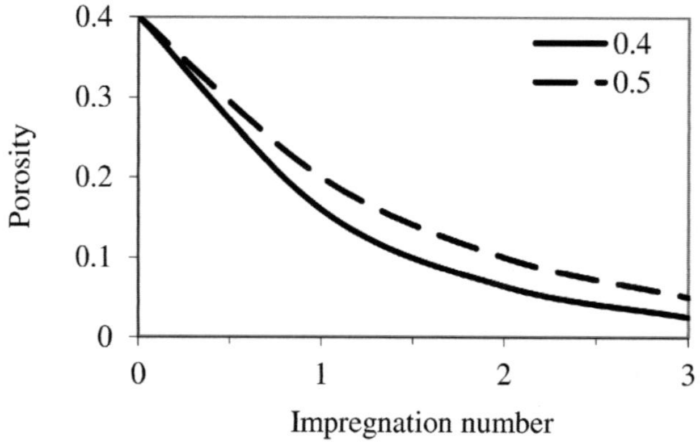

b

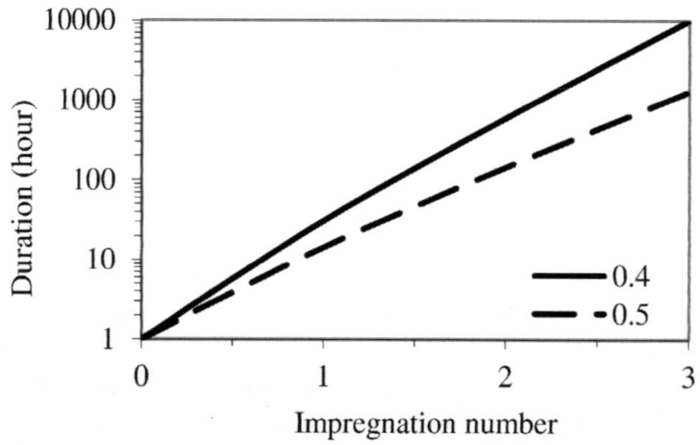

Figure 4. Changes of porosity (a) and (b) binder impregnation process duration of the woven workpiece depending on impregnation number and the percentage of alcohol in the binder solution.

IOP Conf. Series: Materials Science and Engineering **153** (2016) 012015 doi:10.1088/1757-899X/153/1/012015

As it is shown in experiments with resin B71731AL[12, 13, 21, 23], a reduction of binder viscosity can be expected by the introduction of carbon nanotubes (CNT).

3. The joint processes of binder curing and compacting of composite thermal protection filler in the autoclave

The processes of binder curing and compacting of the composite thermal protection filler are based on the possible uses of the RSCE autoclave.

The autoclave plant (figure 5) is a stand-alone automated system, integrating a thermally insulated vertical high pressure vessel, with a volume of 293 m^3, the diameter of working space is 5.5 m, and the height is 6 m. The autoclave includes also: heating and cooling blocks, gaseous nitrogen compressed to 3.5 MPa production station, the nitrogen cumulative installation with the capacity of 300 m^3, the vacuum station, the water conditioning unit with graduation towers and the chemical treating assembly of pipelines.

Figure 5. Autoclave plant for DS composite structures forming:
1 – Nitrogen cumulative installation; 2 – Nitrogen plant; 3 – Vacuum station; 4 – Autoclave.

Special attention is paid to create conditions for a minimum temperature gradient on products during heating, thermostatic control and cooling, as well as to ensure even distribution of the binder in the material in case of multilayer, large-dimension and complex shape objects.

To achieve this result, the circulation system of coolant laminar flow movement with an optimum orientation relative to the product surface was developed (figure 6). On this plant basis, the thermal protection in neutral environment technology was developed for the transport spaceship "Soyuz TMA". This technology provides reliability and stability of production, thermal protection quality and improved reliability, an increase of labor productivity and the reduction of expense of the electric power.

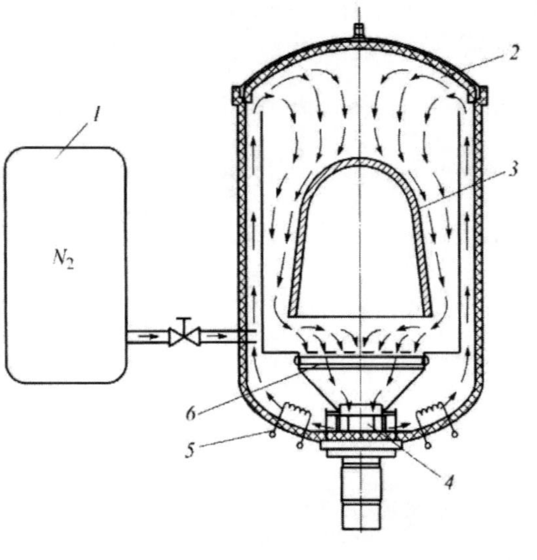

Figure 6. Scheme of nitrogen heating and movement in reaction chamber of autoclave with the following parameters:
1 — receiver;
2 — reaction chamber;
3 — DS thermal protection cover workpiece;
4 — fan;
5 — heater;
6 — heat exchanger

An autoclave control system enables change in pressure and temperature by pre-set programs $p(t)$ and $T(t)$. A control program must be selected to cure the binder and seal the filler structure with a minimum of compressed nitrogen. The program selection procedure is a complex scientific task, which mathematically requires detailed discussion.

The medium used in the autoclave is nitrogen which is obtained from air at the autoclave nitrogen plant and is pressurized by a compressor up to the receiver pressure. If required, nitrogen is supplied from the receiver to the reaction chamber, thus, creating the required pressure. The reaction chamber temperature is changed by nitrogen heating with varied power thermoelectric heaters.

There are five thermoelectric heater circuits in the autoclave, for each control stage of electric power. A fan located at the lower point of the reaction chamber provides heated nitrogen flow between the chamber walls. Near to the cover, on the top of the reaction chamber, the flow turns back to the centrally located product.

The law of product temperature variation with time and thermoelectric heater power during the autoclave treatment is selected on the base of practical experience of the product quality assurance. Thermoelectric heater power and temperature were recorded by a control system (figures 7 and 8) and used at determination of specific features of the autoclave.

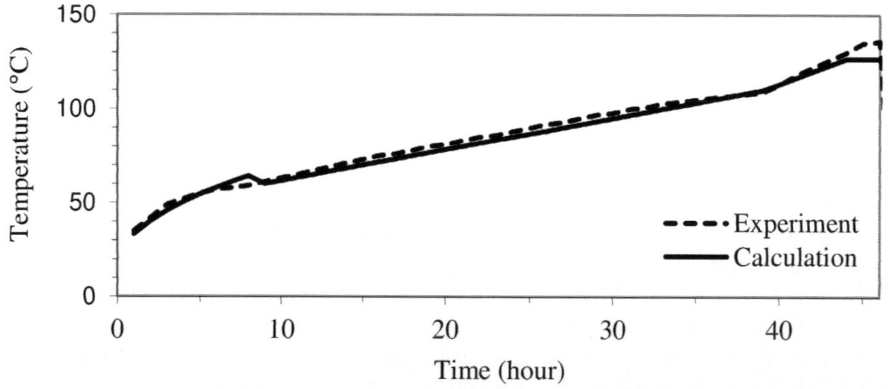

Figure 7. Temperature T variation during heat treatment of the product.

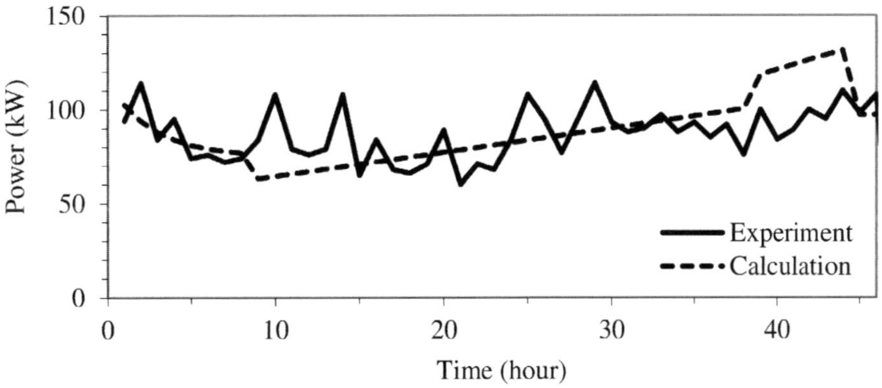

Figure 8. Thermoelectric power W variation during heat treatment of the product.

There are four sections in the temperature variation curve, where temperature variation rate $\dot{T}$ was selected equal to 10, 1.67 and 3.33 K/hour. In order to reduce the time required for heat treatment to eliminate significant deterioration of product quality on the 1st and 3rd sections, the temperature rate was increased compared to the recommended rate on the 2nd section.

The temperature variation rate is defined as follows, $\dot{T} = 1/\ddot{t}\delta$, where δ is the wall thickness of the product, $\ddot{t} = \text{constant}$ is the recommended temperature rise by one degree assigned to the wall thickness unit. The proposed recommendations on the selection of thermoelectric heater power may relate to various structural elements of the thermal protection of descent spaceships.

The correlation of temperature-to-electric power based on the heat-transfer response rate can be described using the following equation

$$\frac{dT}{dt} + \mu T = \chi W \tag{2}$$

where μ is the factor of process response rate of the heat transferred from thermoelectric heat to the product.

The determination of the factors χ and μ is an important stage of establishing the procedure aimed to correlate autoclave nitrogen supply and temperature variation of composite products when the matrix is cured. Based on the data obtained by joint analysis using autoclave control instrumentation (the analysis was conducted on the four heating sections for stepwise thermoelectric heater circuit control), the following factor values were set as $\chi = 0.0972$ K kW^{-1} h^{-1} and $\mu = 0.0745$ h^{-1}.

Nitrogen pressure is varied at the same time when the temperature of the product in the autoclave rises. It enables the combination of the process of binder curing and filler sealing in the product wall, from remains of the binder extruded from the work-piece.

The wall thickness variation law is obtained after the Darcy equation, (1) is re-arranged

$$\frac{d \ln \delta}{dx} = -\frac{2K_k^*}{R^2 + 2RH} \frac{\Delta T_m}{\dot{T}} \frac{\Delta p}{\eta_0^* \eta(x)} \tag{3}$$

where $K_k^* = K_k R/(R + H)$ is a factor which considers the complex geometry of product side surface profile, R and H are the radius and height of the product, η_0^* is the binder viscosity at the initial instant at normal temperature, $x = \Delta T/\Delta T_m$, and ΔT_m is the maximum temperature differential during the autoclave treatment.

The dependence of dimensionless viscosity η on relative temperature x was obtained as an approximation result from experimental data. Thus, the data was obtained for epoxy-based binder ENFB/3692 [18, 22] which is approximated by the following formula

$$\eta = \frac{b\exp(-\beta_T x)}{(1-ax)^n} \tag{4}$$

and as shown in figure 9, where the numerator means the viscosity drop when heated, and the denominator means the viscosity increase as a result of polymerization. a, b and n are empirical factors that have to be defined using figure 4, where the constants are as follows: $a = 0.997$, $b = 0.3$, $n = 1$, and $\beta_T = 4.8283$.

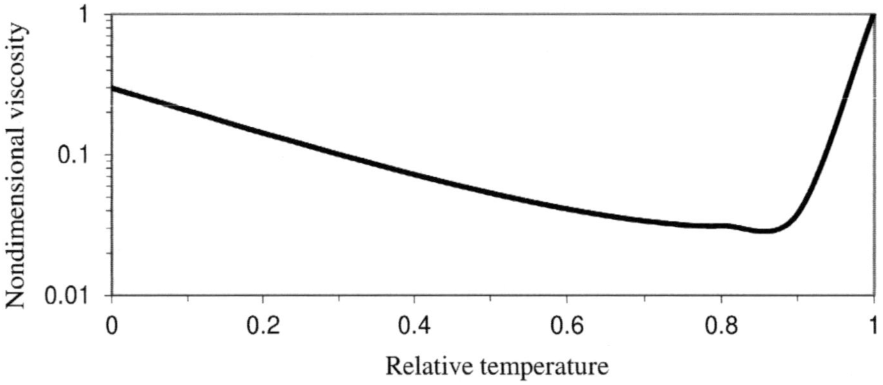

Figure 9. Nondimensional viscosity, η, *vs.* relative temperature x.

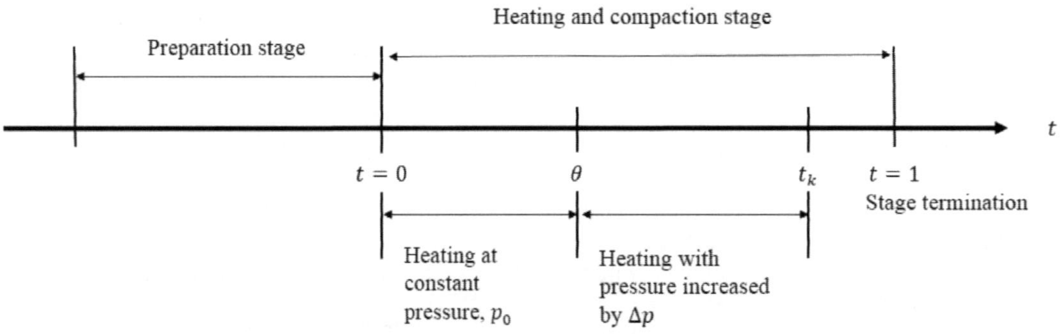

Figure 10. Time-line for the manufacturing process illustrated in nondimensional time.

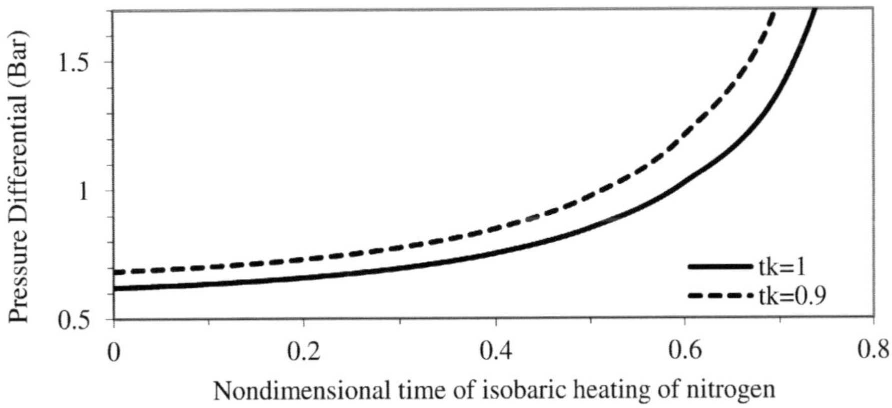

Figure 11. Nitrogen pressure differential amplitude, Δp, vs. nondimensional time, θ, during the isobaric heating of the nitrogen. The quantity tk represents the relationship between the time taken to achieve the required binder volume fraction and the heating of the product.

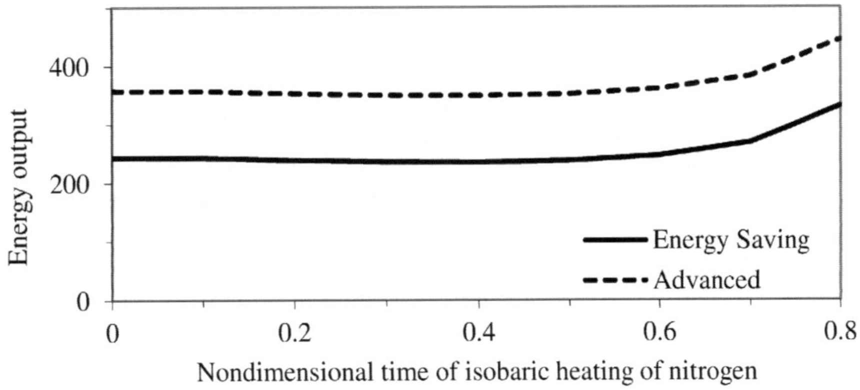

Figure 12. The power cost to obtain the required nitrogen mass vs. nondimensional time, θ, during the isobaric heating of the nitrogen. Continuous and dotted lines correspond to the economic and progressive modes, respectively, of the filling of the autoclave with nitrogen.

After the integration of equation (3), the correlation between the instant of nitrogen supply and its autoclave pressure differential amplitude appears, figure 11. The thermal effect on nitrogen-side thermal protection includes nitrogen refrigeration when it is adiabatically supplied from the receiver, nitrogen heating at first at the constant volume, and then with a constant pressure. The required variation of temperature is provided as a result.

For this purpose the automatic equipment of the autoclave carries out a step change of power of thermoelectric heaters and step supply of nitrogen, as represented in figure 8.

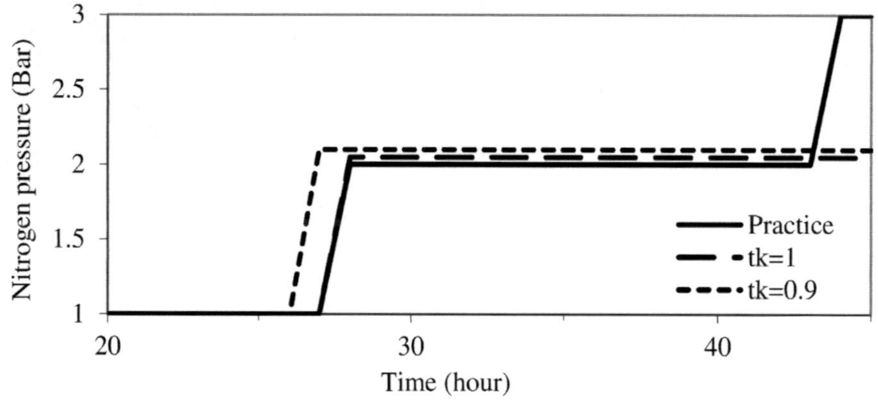

Figure13. Comparison of theoretical and experimental recommendations on pressure Δp in time t. The continuous line represents the practical recommendation, and the dashed lines are theoretical recommendations corresponding to different values of t_{k1} and t_{k2}.

It is convenient to present the key processing events in terms of dimensionless time, t, which is obtained by dividing the actual time by the duration of the heating process, see figure 10.

The production cycle can be carried out in accordance with either the economic or progressive modes. The preparation stage for the operation of the autoclave, shown in negative area of dimensionless time, involves the filling of the autoclave reaction chamber with nitrogen. This stage of the economy mode consists of: creation of vacuum in the chamber; releasing nitrogen, of mass m_0 from the receiver and heating until the pressure reaches $p_0 = 10^5$ Pa at room temperature 20 °C. The first stage of the progressive mode consists of pressurizing the nitrogen in the chamber before reaching the required concentration.

The second stage is identical in both modes. It includes isobaric heating of nitrogen at p_0 until dimensionless time, θ. At this moment the additional mass of nitrogen, Δm, enters the chamber, and electric heaters maintain the temperature. This leads to an increased nitrogen pressure, $p_0 + \Delta p$, which influences the rate of displacement of excess binder volume from the workpiece. The displacement process at uniform heating and constant pressure and comes to the end at a time, $t_k \leq 1$, (≈ 1), when the required fibre-matrix volume fraction in composite material of thermal protection is reached.

In this paper, it is recommended to use variable-based power consumption as a selection criterion for instant of time

$$E_{\Sigma}^{\mathrm{var}} = E_{00} + E_0 + \Delta E_1 + \Delta E_2 + E_1 + E_2 \tag{5}$$

where $E_{00} = m_0 E_{N_2}$ is the power cost to obtain nitrogen mass m_0 from air and to compress it up to receiver pressure, E_{N_2} is the power cost to supply 1 kg of nitrogen from nitrogen plant, fitted in the used autoclave to receiver, E_0 is the power consumption to restore temperature in nitrogen with mass m_0, ΔE_1 is the power cost to obtain nitrogen with mass Δm from air and to compress it up to receiver pressure, ΔE_2 is the power consumption to restore the temperature in nitrogen with mass Δm, E_1 and E_2 are the power consumptions during an even increase of the product temperature before and after nitrogen auxiliary supplied to autoclave, accordingly.

When cost-saving and progressive pressure and temperature control algorithms are implemented, the calculation results of variable-based power consumption are represented in figure 12.

The nondimensional time values are defined in terms of actual times, as follows

$$\theta = \frac{t_S}{t_T} \quad \text{and} \quad t_{ki} = \frac{t_R}{t_T} \tag{6}$$

where t_S is the point in time when the additional mass of nitrogen was supplied, t_R is the time to achieve the required volume fraction, and t_T is the time at which the product heating stage was finished.

Analysis of the calculation result demonstrates that the power consumption at level $\theta = 0.4 - 0.5$ is minimum, therefore, the pressure variation shall be supplied after $\theta > 0.5 - 0.6$. Comparison of the proposed procedure and experimental application of the law of stepwise pressure variation is presented in figure 13.

The compatibility of instant of auxiliary nitrogen supply and nitrogen pressure variation amplitude is observed to be satisfactory. The mathematical models presented in the current work could be used for technological design of new spaceships.

4. Conclusions

A method is presented, demonstrating the effect of process duration and the number of repetitions on the impregnation operation, taking material porosity and binder viscosity into account. The method includes the calculation of the consumed mass of nitrogen, and for the consumption of electrical power for the receiving and heating of the nitrogen. It is shown that, to minimize the electrical power expense, there is a criterion for coordinating the moment of additional supply of nitrogen and the pressure amplitude. Comparison of the results obtained through calculation and experiment indicates that these techniques are highly effective.

References

[1] Polezhayev Yu V and Yurevich F B 1976 *Thermal protection,* ed A V Lykov (Moscow: Energy)

[2] Dushin Yu A 1968 *Work of heat-shielding materials in hot gas streams* (Moscow: Chemistry)

[3] Peterson D L, Nahtsgame P R and Houa J T 1972 Use of the reflecting ablative materials for a heat-shielding of the devices entering the atmosphere of planets *Rocketry and Astronautics* no II pp. 137—145

[4] Kamalov V S 1985 *Production of spacecrafts* (Moscow: Mechanical engineering)

[5] Semenov A A 2009 *The lowered spacecraft capsule* (Sankt Peterburg: Neva)

[6] Vlasov V I, Zalogin G N, Kovalyov R V and Churakov D A 2012 Radiant and convective heat exchange of the lander with the destroyed thermal protection *Physical and Chemical Kinetics in Gas Dynamics,* http://chemphys.edu.ru/media/files/2012-12-26-001.pdf

[7] Tarasov V A, Komkov M A and Romanenkov V A 2014 Технологические основы снижения длительности цикла и повышения безопасности изготовления тепловой защиты спускаемых космических аппаратов [Technological bases of decrease in duration of a cycle and increase of safety of production of thermal protection of the lowered spacecrafts] *Proc Higher Educational Institutions, Machine Building* no 8 pp 35—43

[8] Strekalov A F, Pashchenko V A, Romanenkov V A, Morokova E V, Bazanov V V, Zimovsky A V, Andriyanov V S, Starostin V V, Tarasov V A and Filimonov A S 2010 *Thermal and vacuum equipment for processing of a product(s)* Patent No. 2439455 Russian Federation bulletin no 26

[9] Strekalov A F, Pashchenko V A, Romanenkov V A, Morokova E V, Bazanov V V, Zimovsky A V, Andriyanov V S, Starostin V V, Tarasov V A and Filimonov A S 2010

Method of production of multilayered products Patent no. 2450921 Russian Federation bulletin no 26

[10] Tarasov V A, Romanenkov V A, Komkov M A and Boyarskaya R V 2015 Методика выбора режимов термобарической обработки в автоклаве теплозащиты спускаемых космических аппаратов [Technique of a choice of the modes of descent spacecrafts thermal protection in the autoclave of a heat-shielding of the lowered spacecrafts] *Proc Higher Educational Institutions, Machine Building* no 9 pp 83—92

[11] Komkov M A and Tarasov V A 2014 Influence of viscosity binding in an impregnating bathtub on porosity of a composite at a wet way of winding *Science and education* no 12 pp 192—199, DOI: 10.7463/1214.0745284

[12] Tarasov V A, Stepanishchev N A, Romanenkov V A and Alyamovsky A I 2012 Improvement of quality and technological effectiveness of a polyester matrix of composite designs on the basis of ultrasonic nano modifying *Herald of the Bauman Moscow State Technical University. Series "Mechanical Engineering"* pp 166—174

[13] Tarasov V A, Komkov M A, Stepanishchev N A, Romanenkov V A and Boyarskaya R V 2015 Modification of Polyester Resin Binder by Carbon Nanotubes Using Ultrasonic Dispersion *Polymer Science, Series D* vol 8 no 1 pp 9–16, DOI: 10.1134/S1995421215010165

[14] Komkov M A 2012 Rheological properties of the polymeric binding, used when winding products from composite materials *Glues Sealants Technologies* no 3 pp 30—35

[15] Komkov M A, Tarasov V A and Kuznetsov V M 2015 Research of influence of viscosity of epoxy fibrous fillers, binding on extent of impregnation *Glues Sealants Technologies* no 4 pp 24—27

[16] Komkov M A 2013 Rheological Properties of Polymer Binders Used in the Windings of Products of Composite Materials *Polymer Science Series D* Vol 6 Issue 1 pp 26—30, DOI: 10.1134/S1995421212040089.http://link.springer.com/article

[17] Komkov M A and Tarasov V A 2011 *Technology of winding of composite designs of rockets and means of defeat* (Moscow: Publishing house of BMSTU)

[18] Tarasov V A and Belyakov E V 2011 Mathematical modeling of process of not isothermal hardening of polymeric composite designs of RKT *Herald of the Bauman Moscow State Technical University. Series "Mechanical Engineering"* no 1(82) pp 106—116.

[19] Tarasov V A and Belyakov E V 2011 Dynamics of consolidation of a composite at a hot pressing *Proc Higher Educational Institutions, Machine Building* no 12 pp 53—57

[20] Tarasov V A, Belyakov E V and Boyarskaya R V 2012 Choice of the modes of formation of composite designs of the missile and space equipment *Proc Higher Educational Institutions. Machine Building* no 5 pp 37—43

[21] Stepanischev N A and Tarasov V A 2011 Increase of durability of polymer matrix composite materials based on nanomodification of binding *Proc First International Workshop on Advanced Composite Materials and Technologies for Aerospace Applications* Wrexham UK pp 53—60

[22] Belyakov E V, Boyarskaya R V and Tarasov V A 2011 Viscosity and kinetics of epoxy of composite constructions manufactured by method of vacuum forming *Proc First International Workshop on Advanced Composite Materials and Technologies for Aerospace Applications* Wrexham UK pp 121—127

[23] Tarasov V A, Stepanischev N A and Boyarskaya R V 2012 An Experimental Method for Determination of Characteristic Time Points in Nanosuspensions Processing under Influence of Ultrasonic Field *Proc Second International Workshop on Advanced Composite Materials and Technologies for Aerospace Applications* Wrexham UK pp 34—37

[24] Malysheva G V, Akhmetova E Sh and Marycheva A N 2014 Assessment of vitrification temperature epoxy binding *Physics and Chemistry of glasses* vol 40 no 5 pp 718—724

[25] Collins R E 1961 *Flow of fluids through porous materials*, (Reinhold Pub. Corp.)

IOP Conf. Series: Materials Science and Engineering **153** (2016) 012016 doi:10.1088/1757-899X/153/1/012016

Damage monitoring of aircraft structures made of composite materials using wavelet transforms

Molchanov D[1], Safin A[1], Luhyna N[2]

[1] Samara State Aerospace University, 34, Moskovskoye Shosse, Samara, 443086, Russia

[2] Advanced Composite Training and Development Centre, School of Applied Science, Computer Science and Engineering, Wrexham Glyndŵr University, Mold Road, LL11 2AW, Wrexham, UK

molchanovdmitriy@icloud.com

Abstract. The present article is dedicated to the study of the acoustic properties of composite materials and the application of non-destructive testing methods to aircraft components. A mathematical model of a wavelet transformed signal is presented. The main acoustic (vibration) properties of different composite material structures were researched. Multiple vibration parameter dependencies on the noise reduction factor were derived. The main steps of a research procedure and new method algorithm are presented. The data obtained was compared with the data from a three dimensional laser-Doppler scanning vibrometer, to validate the results. The new technique was tested in the laboratory and on civil aircraft at a training airfield.

1. Introduction

As a result of a significant growth in the use of composite materials in aviation structural components there is a strong need for increased capability for the operational monitoring and inspection of such components. The main impact of this research lies in the management and operation of future composite aircraft components. The data, which was obtained, can help to develop a universal non-destructive testing method, and as a result, improve aviation safety. The condition monitoring of modern passenger aircraft with complex composite material components, and the prediction of the operational state of such components, will be one of the major aviation operational tasks of the next decades.

During the preliminary research it was concluded that wavelet analysis is the most effective and appropriate tool for the vibration analysis for the determination of the modal parameters of free oscillations, especially in the integral control of a composite structural aircraft component, during the operational inspection. Wavelets can be combined, using a "reverse, shift, multiply and integrate" technique called convolution, with portions of a known signal to extract information from the unknown signal. The present research represents a "shift" technique with the use of the variable list of the wavelet family, such as Morlet, Gabor and Daubechies wavelets [1].

Other diagnostic and NDT methods, materials and results have been studied in detail to verify the correctness of the chosen research method [2-8].

The aircraft components of interest in the research were: propeller blades (figure 1); Yak-42 aircraft spoilers (figure. 2); and different types of honeycomb construction.

Figure 1. Propeller blades with different operation time.

Propeller blades with different surface conditions and geometry were tested. The fully composite blade with an Aluminum alloy ring at the attachment point is a perfect example of a modern hybrid composite structure. Each blade was made from pre-preg material using different types of epoxy resin. These blades are commonly used in small civil aviation aircraft, especially in hydro-aeroplanes, trikes, trainer aircrafts, etc.. The life of each blade, according to passport data, is, on average, no more than 600 flight hours. Yak-42 aircraft spoilers, made from a three layer reinforced honeycomb construction, were also tested in lab and on the training airfield.

2. Theoretical basis

The wavelet transform can be regarded as a three-dimensional spectrum, where time is given on the X – axis, frequency on the Y – axis, and the Z – axis represents the amplitude of harmonics with a given frequency at a given time. There are two main features that make the continuous wavelet transform (CWT) particularly attractive. First, in most cases, the vibration modes are automatically separated, thus providing the natural frequencies. This means that the values of the instantaneous frequencies and damping parameters can be easily obtained. Second, the most significant information is preserved in the CWT line highs and ridges (the high amplitude areas of the CWT figure). It should be noted that for the appropriate choice of the wavelet family, CWT wavelets can serve as a filter for separation of free oscillations.

To illustrate the proposed method, we analyze a simple case of a linear system with two degrees of freedom (figure 3). This shows the principle for obtaining instantaneous frequency and damping parameters through the use of CWT (calculated only for ranges).

$$f(t) = \sum_{j=1}^{2} A_j(t) cos\Phi_j(t), \tag{1}$$

where $A_j(t) = \alpha_j \exp\left(-\zeta_{j\omega_j} t\right)$, and $\Phi_j(t) = (1-\zeta_j^2)^{\frac{1}{2}}\,\omega_j t + \varphi_j$ and ω_j are the natural oscillation frequencies, with damping coefficient, ζ_j; initial amplitude, α_j ; and the phase shift of the jth mode, φ_j.

Table 1. Values used in the linear system example

α_1=	α_2	ζ_1	ζ_2	ω_1	ω_2	φ_1	φ_2
0.5	3.0	0.03	0.045	40 π	156 π	0.5 π	0.5 π

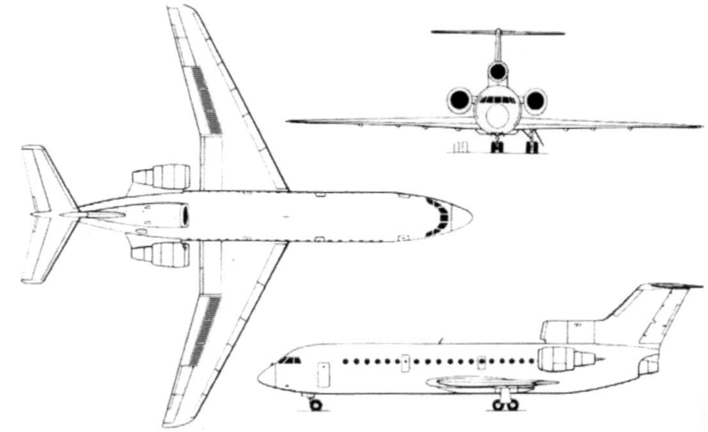

(a)

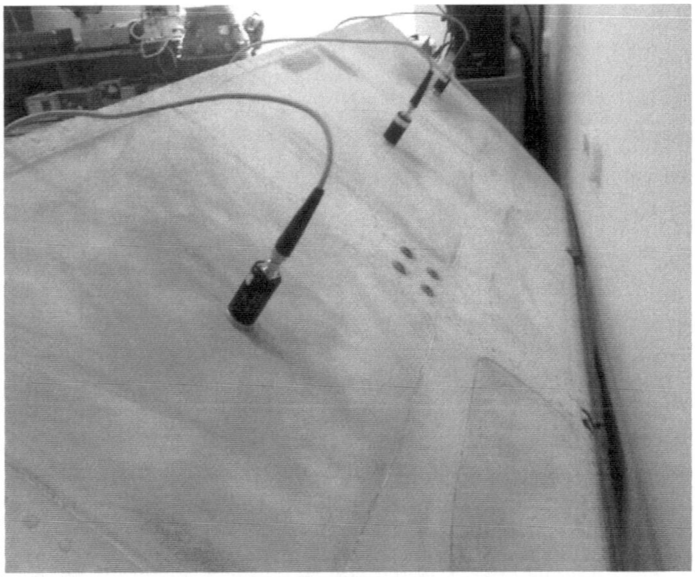

(b)

Figure 2. (a)Yak-42 passenger jet with spoilers shown in red;
(b) Yak-42 spoilers in lab during the testing.

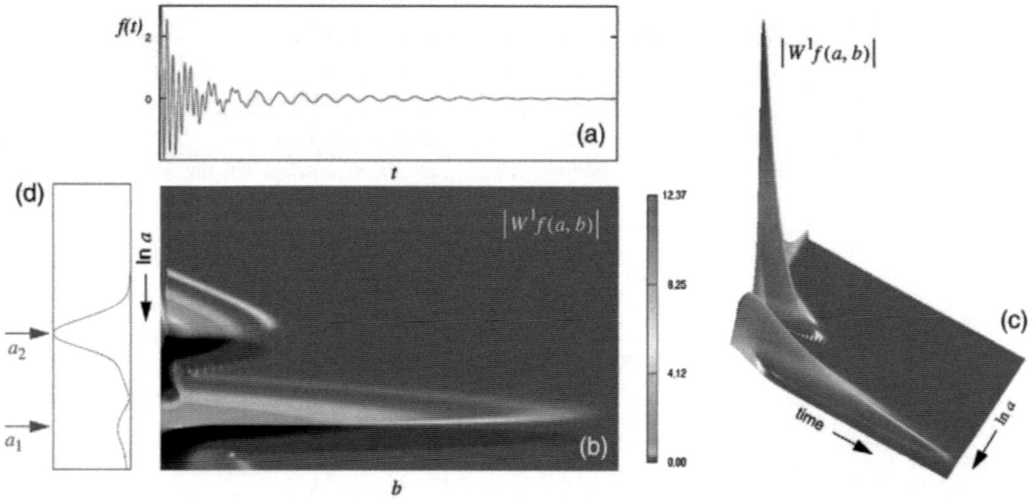

Figure 3. Linear system with two degrees of freedom: (a) Function *f(t)*; (b, c) Wavelet transform; (d) Free oscillation frequency.

Figure 3 (a) shows the reaction to the impulse excitation during the time interval ($0 \leq t \leq 1$). The first-order derivative for the CWT wavelet analysis was selected to be equal to

$$\psi_1(t) = (-t + i\omega_0)e^{\frac{-t^2}{2}}e^{-i\omega_0 t}, \quad (2)$$

with $\omega_0 = 5$. Figures 3 (b) and (c) show two a 2D and a 3D view of the CWT, from which the modes of oscillation can be clearly identified; however, the proposed method does not require a complete calculation of the CWT. It is only necessary to perform the calculation for a fixed time, b_0, figure 3 (d), and evaluate the scales a_1 and a_2 for the maximum deformation: $a_1 = 4.159\text{x}10^{-4}$; $a_2 = 1.046\text{x}10^{-2}$. The relevant frequencies, ω_j, and damping parameters, ζ_j, can be determined by the following formula [9]

$$a_j\sqrt{1-\zeta_j^2}\omega_j = \omega^* \text{ and } \zeta_j\omega_j = -m_j. \quad (3)$$

3. Methodology and experimental results

The data obtained was validated by comparison with data from a Polytec PSV-400 scanning Vibrometer (figure 4). The method of the procedure contains several iterations combining direct Fourier transformation and CWT, (figure 5).

Using a direct Fourier transform – first iteration, (figure 6), to check the dependence of the vibratory acceleration on the frequency, can be used to indicate the presence of a defect.

Figures 7, 8 and 9 illustrate the features of the CWT before and after the appearance of the artificially created defect in the reference honeycomb structure. There is a clear difference between the CWT result for the object before and after the appearance of the defect. The differences marked in yellow highlight the changes in particular areas.

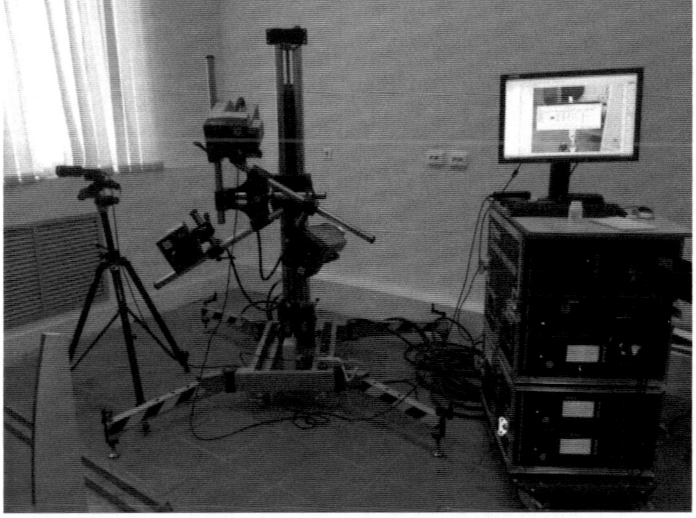

Figure 4. Polytec PSV-400 scanning Vibrometer.

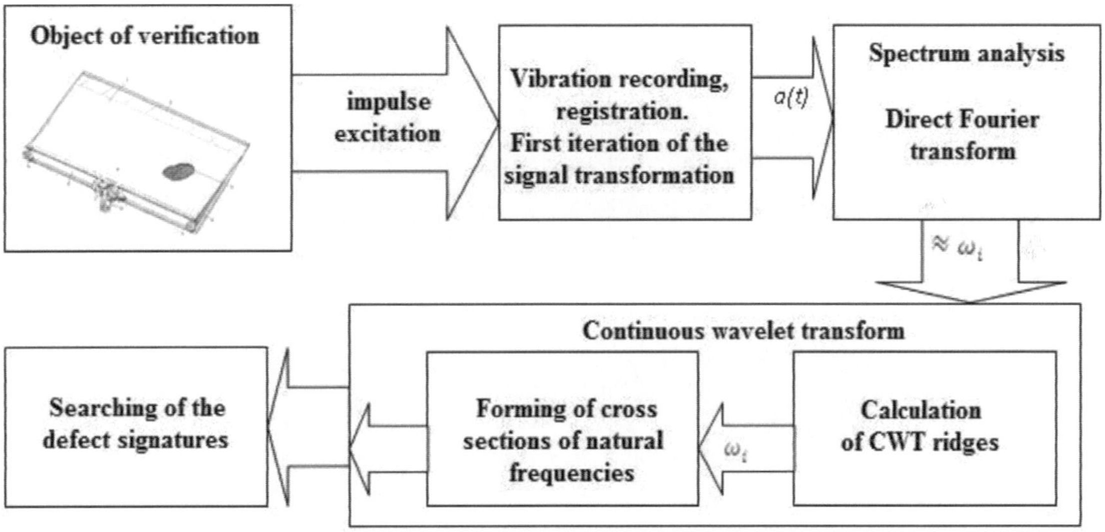

Figure 5. The main steps of the procedure (Algorithm).

The area of the defect in the reference honeycomb construction was gradually increased, from approximately 1% to 5% of the internal honeycomb construction space. To create the defect, water was injected water, and subsequently frozen, to cause the delamination between the honeycomb core and the layers. After that, the water was vaporized, and the sample was weighed to ensure that there was no remaining water in the construction.

The works [10-14] guided the authors in the interpretation of the CWT signal images.

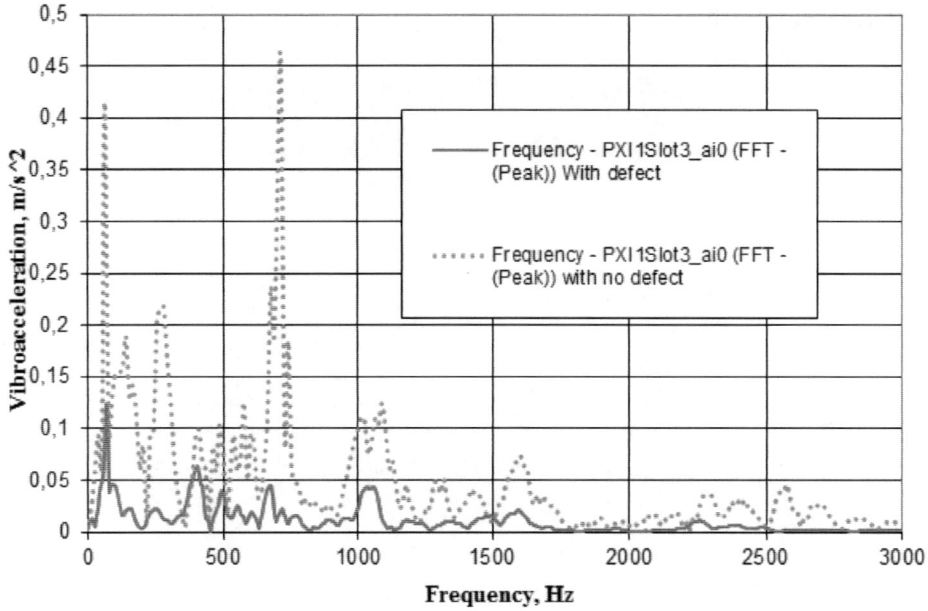

Figure 6. Graph of Vibratory acceleration, with and without the defect.

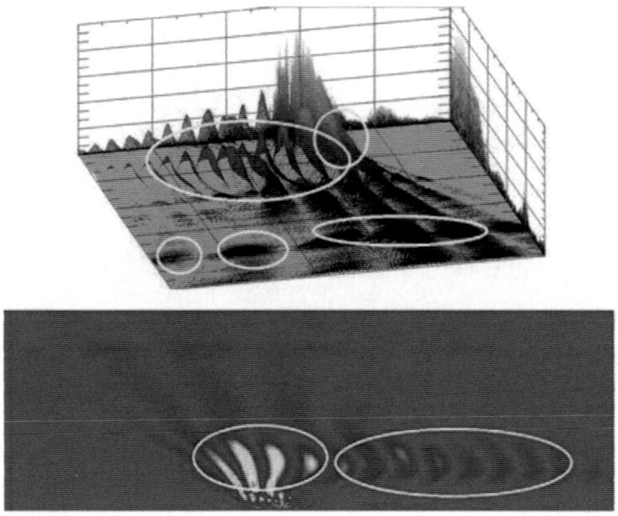

Figure 7. CWT 3d and Scalogram – honeycomb construction with no defect.

4. Discussion

As might be anticipated, the new testing experimental procedure, carried out on the reference honeycomb structure, showed a reduction in the value of each natural frequency and an increase in damping at resonance as a result of the presence of damage. Visual features of the CWT 3d and scalogram indicate changes in the overall vibration picture of the signal, which can directly attributed to the integral damage condition of the object under investigation.

The aircraft components for which this procedure would be most applicable include spoilers, propeller blades and different types of honeycomb constructions.

It was found that the master blade mode shapes at resonant frequencies are practically the same in comparison with a blade following an operating time of 500-600 flight hours (figure 10).

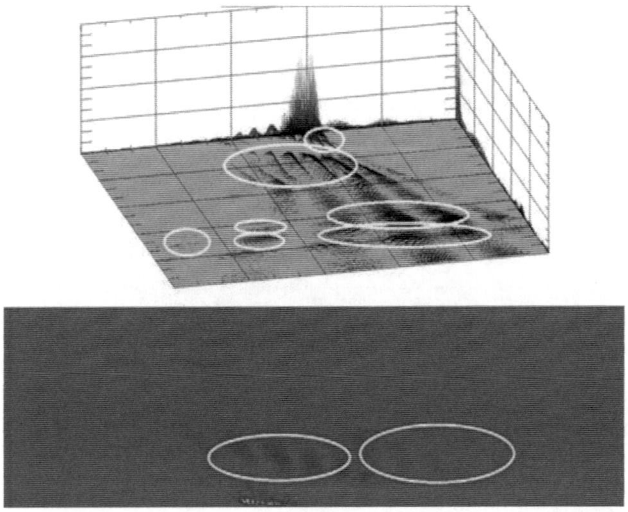

Figure 8. CWT 3d and Scalogram, honeycomb with 1% artificially created delamination.

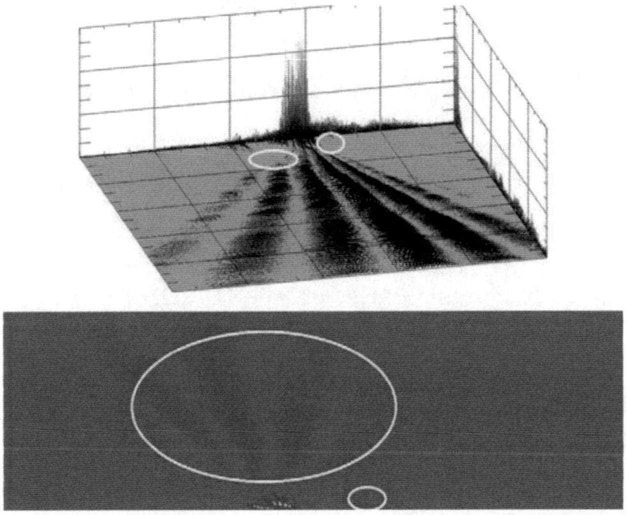

Figure 9. CWT 3d and Scalogram, honeycomb with 5% artificially created delamination.

Vibration speed and vibratory displacement are presented (figure 11). It is shown that there is an increase in the vibration speed on some certain frequencies.

5. Conclusion

This research indicates that wavelet based NDT analysis could provide a basis for determining damage levels in structural composite aerospace components, and thus a means to decide whether a component is still fit to be operational.

IOP Conf. Series: Materials Science and Engineering **153** (2016) 012016 doi:10.1088/1757-899X/153/1/012016

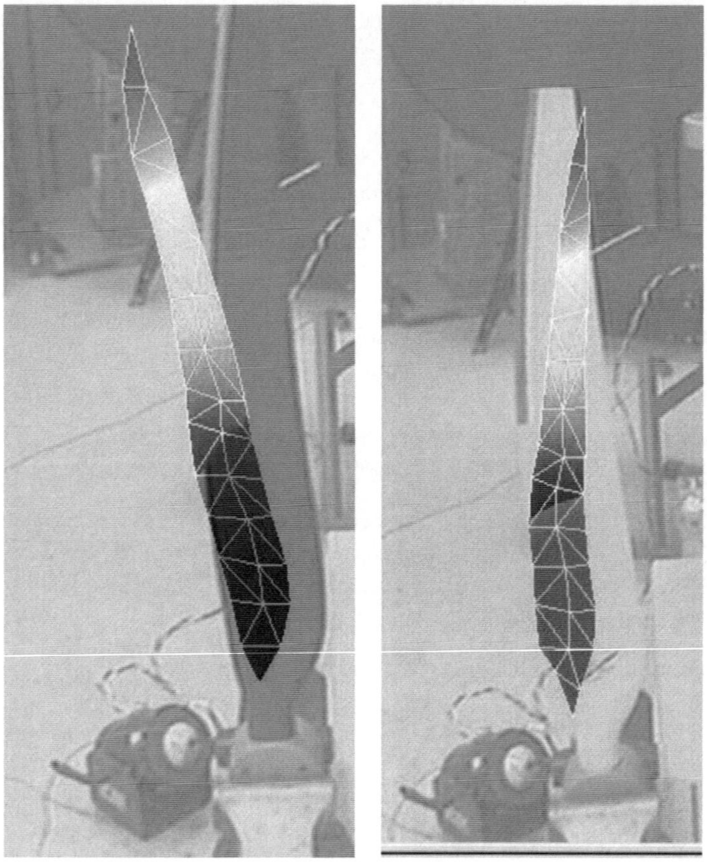

(a) (b)

Figure 10. Blade mode shapes obtained using a 3 dimensional laser-Doppler scanning vibrometer
(a) Master blade
(b) Blade with 500-600 flight hours.

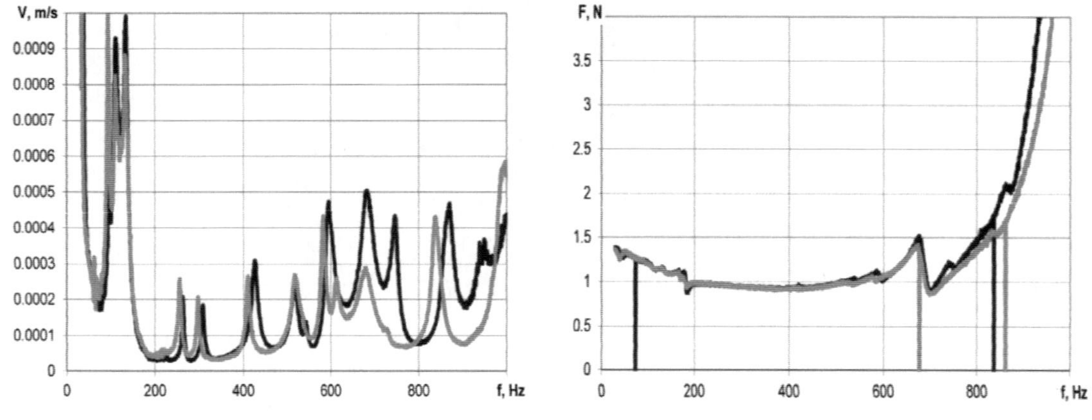

(a) (b)

Figure 11. Comparison of master blade (pink) and blade after 500-600 hours operation (blue).
(a) Vibration speed and (b) Vibratory displacement.

The method presented has the potential for critical life-cycle parameter identification, with a view to enabling life extension certification, based on acoustical structure health monitoring. It should be mentioned that hardware and software can be used both for searching for defect signatures, and also for structural control. The data obtained during the test series indicated that the maintenance process is sufficiently frequent for predicting time to failure based on observed damage.

References

[1] Daubechies I (1992) *Ten lectures on wavelets*, University of Lowell, Philadelphia: Society for Industrial and Applied Mathematics (SIAM)
DOI: http://dx.doi.org/10.1137/1.9781611970104

[2] Blanas P and Das-Gupta DK (2000) Composite piezoelectric materials for health monitoring of composite structures. *Materials Research Society Symposium Proceedings*. Materials for Smart Systems III: Symposium, 30 Nov–2 Dec 1999, Boston, MA, USA, **604** 45–50

[3] Koktavy P, Trcka T and Koktavy B (2011) Noise diagnostics of advanced composite materials for structural applications. *Proceedings IEEE 21st Int Conf Noise and Fluctuations*, 84–7
DOI: 10.1109/ICNF.2011.5994391

[4] Makeev A and Nikishkov Y (2011) Material characterization and failure prediction for composites. *37th European Rotorcraft Forum*, 13–15 Sept 2011, Vergiate and Gallarate, Italy, **2**, 1164–70

[5] Stifter D, Wiesauer K, Wurm M, Schlotthauer E, Kastner J, Pircher M, Götzinger E and Hitzenberger CK (2008) Investigation of polymer and polymer/fibre composite materials with optical coherence tomography. *Measurement Science and Technology*, **19**(7)
DOI: 10.1088/0957-0233/19/7/074011

[6] Ecault R, Boustie M, Touchard F, Berthe L, Chocinski-Arnault L, Ehrhart B and Bockenheimer C (2011) Damage of composite materials by use of laser driven shock waves. *2nd Joint US–Canada Conf Composites.* Montreal, Quebec, Canada, 26–28 September 2011, **3**, 2583–96

[7] Ben BS, Ben BA, Vikram KA and Yang SH (2013) Damage identification in composite materials using ultrasonic based Lamb wave method. *Measurement*, **46**(2), 904–12

[8] Karabutov AA, Murashov VV, Podymova NB and Oraevsky AA (1998) Nondestructive characterization of layered composite materials with a laser optoacoustic sensor. *Proceedings* SPIE, San Antonio TX, ETATS-UNIS, 31 March 1998, **3396**, 103–11

[9] Haase M and Widjajakusuma J (2003) Damage identification based on ridges and maxima lines of the wavelet transform. *Int J Engineering Science*, **41**, 1423–43

[10] Destic F, Petitjean Y, Mollier J, Barbieri S and Sirtori C (2009) THz QCL - based active imaging applied to composite materials diagnostic. *34th Int Conf Infrared, Millimeter, and Terahertz Waves, IRMMW-THz*, Busan, Korea, 21–25 September 2009, 674–5

[11] Kolodii BI and Lyaschuk OB (1995) Direct and inverse problems of electromagnetic diagnostics of flat-laminated composite materials with the use of a rectangular waveguide. *Materials Science*, **30**(2), 173–81

[12] Kim F, Penumadu D, Siriruk A, Bunn J and Kardjilov N (2011) Three dimensional microstructure of polymeric composite materials used in sandwich structures using dual modality from combined high resolution X-ray and neutron tomography. *18th Int Conf Composite Materials*, Jeju Island, Korea, 21–26 August 2011, 1–6

[13] Avdelidis NP, Moropoulou A and Delegou ET (2003) Applications of thermography in the assessment of masonry, airport pavements and composite materials. *Insight: Non-Destructive Testing and Condition Monitoring,* **45**(12), 836–40

[14] Watson CC, Rhoads JF and Adams DE (2013) Structural dynamic imaging through interfaces using piezoelectric actuation and laser vibrometry for diagnosing the mechanical properties of composite materials. *ASME 2013 Dynamic Systems and Control Conference*, Palo Alto, California, 21–23 Oct 2013, 1–10

IOP Conf. Series: Materials Science and Engineering **153** (2016) 012017 doi:10.1088/1757-899X/153/1/012017

The influence of nano-ceramic modifier on the structure and properties of polyolefin composites

L. Sudnik[1], V. Dubkova[1], A. Galinovsky[2], A. Osipkov[2]

[1] SNU Institute of Powder Metallurgy of National Academy of Sciences of Belarus, Platonova st., 41, Minsk, Belarus

[2] Bauman Moscow State Technical University, 2 Baumanskaya st., 5/1, Moscow, Russian Federation

osipkov@bmstu.ru

Abstract. It has been shown, that the nano-ceramic modifier boehmite, with specific surface area of up to 400 m^2/g and dispersion degree of less than100 nm, is a structurally active filler of an ultra-high-molecular-weight polyethylene. The boehmite influences the polymer crystalline and supra-molecular structure, during crystallization of the polymer from the melt under uniaxial plastic deformation conditions. It results in the production of nano-composites with an improved complex of properties.

1. Introduction

Nowadays, composite materials based on polyolefins are high on the list of thermoplastic materials that are filled and enhanced by various modifiers. The most popular in various technical fields and national economy is ultra-high-molecular-weight polyethylene (UHMWPE) [1, 2]. UHMWPE is distinct from conventional polyethylene in that its molecule consists of hundreds of thousands of units, and the resulting molecular mass is about few millions (> 1.5×10^6 g/mol). Super high molecular mass of this polymer determines its unique properties, which are vastly different from the other types of polyethylene [3–6]. UHMWPE has improved physical and mechanical properties, corrosive and chemical resistance, cold-resistance, extremely high impact resilience, hardness and dimensional stability, working efficiency in a wide temperature range, and resistance to abrasion and gamma-ray radiation. UHMWPE is used in conditions where traditional types of polyethylene and many other polymers do not withstand the severe operating conditions. UHMWPE may be used as an alternative of the other more expensive materials (steel, bronze, polymeric amide, fluorine plastic, etc.) and may be technically required, *i.e.* being the only material with the capability for this purpose.

Industrially, for the first time the UHMWPE was used in 1950s by the Ruhrchemie AG company. In the 1960s AJ Pennings, in co-operation with the DSM company (Netherlands), synthesized fibre structures with extremely high strength from UHMWPE [7, 8]. Currently UHMWPE is produced using various catalysts by a number of manufacturing companies in Japan, USA and Europe; such as Farbwerke Hoechst Ruhrchemie, Ticona (Germany), Mitsui Chemicals Group (Japan), Koninklijke Philips N.V. (Netherlands), [9]. Manufacturing of UHMWPE in the Russian Federation is mastered by PJSC Kazanorgsintez and LLC Tomskiy neftekhimicheskiy zavod.

The properties of the polymer produced depend on the composition of the catalysts and on the ethylene polymerization conditions at these catalysts. Depending on the field of application and

UHMWPE processing methods, various powder types of this polymer are used, which are distinguished by molecular mass, particle size and supramolecular structure. By making appropriate choices, it is possible to change persistently the functional properties of the UHMWPE, and expand the fields of application its composites. Thus, one of the ways to improve the mechanical characteristics of UHMWPE is the addition into polymer of short basaltic fibres or dispersive components as the micro- and nano-particles (kaolinite, carbon nano-tubes, ceramics) [10, 11]. Much attention is given to the development of composite materials based on the UHMWPE with improved tribo-technical properties for operation under dry friction and abrasive wear conditions [12–15].

2. Objective of the work

In the references mentioned above, the functional additives in the UHMWPE compositions are used in limited amounts. The amount of added modifiers is predominantly of 0.5 – 20 % by mass (with a maximum of 40 % by mass) [16]; however, under the conditions of ever higher demand for polymeric composite materials, and the ever increasing scarcity and high cost of the raw material resources that are necessary for polymer manufacture, the objective to reduce the polymeric fraction in compositions while preserving or increasing the specified technical and operational characteristics, is relevant.

The objective of the present investigation is the development and examination of the polyolefin composite properties based on the ultra-high-molecular-weight polyethylene (UHMWPE) and nano-ceramic modifier with increased content of UHMWPE.

3. Experiment

The reactor powder of UHMWPE, produced by LLC Tomskiy neftekhimicheskiy zavod, was used as the object of investigation. The average polymer molecular mass was greater than $4{,}0x10^6$ g/mol. The nano-ceramic modifier choosen was the nano-crystalline aluminum monohydroxide (boehmite), with specific surface up to 400 m^2/g, dispersion degree less than 100 nm. The nano-crystalline aluminium monohydroxide (boehmite), with a specific surface of up to 400 m^2/g and dispersion degree of less than 100 nm, was used as the nano-ceramic modifier. The boehmite was produced by hydrothermal synthesis, in the Belarusian State Research and Production Concern of Powder Metallurgy [17]. Polymeric nano-composites were produced by means of powder technology: dry blend compositions using mechanical activator were pre-cooked, then the blend compositions were mono-lithiated by the method of compression moulding at optimal temperature and time conditions. To compare data results correctly, a UHMWPE sample without the modifying additive was produced, under the same conditions of pre-cooking and moulding.

To examine the structure and properties of the composite, the following methods were used: scanning electron microscopy, electro-physical, acoustic, X-ray phase and thermo-mechanical analyses. X-ray analysis of the boehmite, unfilled UHMWPE, and its compositions with the boehmite after moulding, was carried out on the DRON-3 diffractometer using CuKa-irradiation. Primary processing of the diffraction data obtained, and phase identification of the samples of crystalline compound mixture, were performed using the software packages Powder X and WinXpow [18], and the radiographic powder standards data bank JCPDS PDF2 [19]. Crystallite size analysis was carried out by means of "Size/Strain" software, in order to determine the sizes of coherent scattering region (CSR) and microstresses. In both cases, the calculations are based on integral width variation of diffraction maxima of a crystalline compound with respect to the corresponding line values lying in the same ranges of $2q$ for a standard substance. Here, q is the Bragg angle for diffraction peaks, and the quantity $2q$ is used to determine the inter-planar distances.

A silicon wafer was used as a standard. The degree of crystallinity in the samples was determined by the X-ray diffraction method and using Crystallinity software. The calculations in the software are performed by least square method for all points obtained from the three X-ray diffraction patterns, including the error caused by the air scattering. The patterns include the X-ray diffraction pattern of the substance under investigation and the X-ray diffraction patterns of compounds with known crystallinity indices, which have diffraction maxima in the same 2q region.

Thermo-mechanical investigations of samples, with diameter and height of 10 mm and exactly parallel planes, were performed under heating with constant rate (5 deg/min) using a Hoppler consistometer, modernized for this purpose, under conditions of uniaxial compression and constant pressure of 0.6 MPa. Electro-physical characteristics were determined by means of a E7-14 LCR meter on the operating frequency of 100 Hz, with the error no more than 0.1%. The dynamic modulus of elasticity of the composite was calculated based on the speed of ultrasound transmission through the sample. The hardness of the composite samples was determined on the Hoppler consistometer with the vertex angle of a hardened steel cone of 53° + 10′ on the samples. All sample tests were performed at room temperature.

Morphology of the fractures of UHMWPE compositions with the boehmite was researched by means of a Mira scanning electron microscope with a high resolution, made by Tescan ORSAY HOLDING, a.s. (Czech Republic). The elemental composition was determined by means of energy-dispersive X-ray microanalysis on an INCA 350 test machine, Oxford Instruments (England).

The thermomechanical reseches for the moulded samples of UHMWPE, and its compositions with the boehmite, was determined according to GOST 15088-83 [20]. The essence of the method is the determination of the temperature at which a standard indenter penetrates to a depth of 1 mm, under the action of force, into the test sample heated with the constant speed.

A burning test of a moulded samples of both unfilled and boehmite filled UHMWPE was performed according to GOST 28157-89, Method b [21].

4. Research results

X-ray analysis of tableted UHMWPE, and its compositions with the boehmite, showed that the major crystalline phase of UHMWPE ordered regions in all of the samples is β-polyethylene, which is crystallized in orthorhombic syngony. Along with the crystal peaks of the orthorhombic symmetry group, the peaks on the X-ray diagrams with a lower intensity, related to the monoclinic lattice, were also observed.

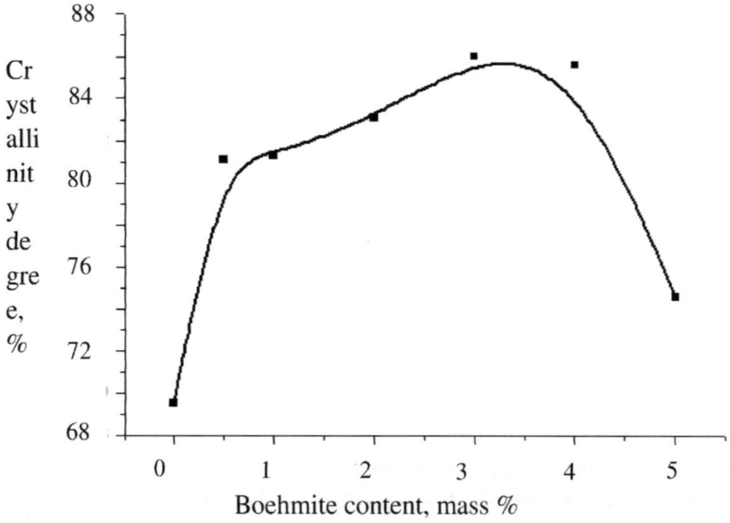

Figure 1. Dependence of X-ray crystallinity degree of UHMWPE on the boehmite content in moulded composites.

Even weak mechanical impacts onto the UHMWPE are shown to disarrange its crystalline structure. The X-ray observed degree of crystallinity is negligibly decreased (up to 1,8 %). It was observed that the monoclinic syngony content increased following mechanical impact; however, after moulding, these changes are levelled, and the monoclinic syngony in unfilled monolithiated UHMWPE is practically non-existent. On the addition of small amounts of the boehmite into the UHMWPE , the monoclinic syngony of UHMWPE crystalline phase is observed again, and the X-ray polymer crystallinity degree is increased. At the same time, this dependence has an extremum (figure 1).

An increase in the dimensions of the crystallite coherent scattering regions (CSR), for the compositions with the boehmite is observed after pressing of the mixed powder structures. At the same time, the crystallites sizes are increased with the degree of increased boehmite in the composition; however, the micro-stresses in the crystalline lattice ($\Delta d/d \cdot 10^{-3}$) are decreased (table 1). The CSR dimension increase of nano-modifier after UHMWPE, with the boehmite powder composition structures moulding, is also observed properly.

Table 1. UHMWPE and Boehmite crystallographic data in moulded compositions.

Boehmite content, % by mass.	CSR dimensions, Å		$\Delta d / d \cdot 10^{-3}$, kg/mm^2	
	Boehmite	UHMWPE	Boehmite	UHMWPE
-	–	252.4	–	5.87
5	171.6	276.9	5.40	5.40
70	455.0	328.2	2.77	4.49

It is noteworthy, that in the X-ray diffraction patterns of the moulded composites of the boehmite with the UHMWPE, new reflections are observed. These reflections point to the structural transformations of the boehmite when composition processing. The reflections are visualized on the fullest extent at mild nano-modifier contents. For instance, an addition of 5 % by mass of boehmite into the UHMWPE composition and mono-lithiating of the composition, results in the observation of reflections related to the crystalline boehmite phase and reflections relating to oxide and aluminium hydroxide. The proportion of the crystalline phases is 54: 35: 11 % by mass respectively. When increasing the boehmite content in the composition to 70 % by mass, the boehmite phase is increased and the percentage proportion is 87: 9: 4 % by mass respectively.

Investigations of the supramolecular structure of the various compositions by electron microscopy showed the following results. On the spalled sample of the moulded unfilled UHMWPE, spots of a prolate spherilitic structures with the length up to dozens μm are observed. The spherulites are oriented by their extended sides in the plane perpendicular to the pressure direction during moulding, see figure 2(a). The spherulites are distributed uniformly enough on the entire spalled spot surface and interstratified with amorphous (disarranged) component, figure 2(b), concentrated at the outer surfaces of the polycrystalline aggregates. When UHMWPE crystallizes from the melt with the boehmite under uniaxial plastic deformation conditions, the general character of the morphological formation is not changed. A formation of spherulites also takes place; however, there is a tendency to its increasing, figure 2(c). The nano-modifier was displaced in an amorphous polymeric phase, incorporated in its supramolecular structures, which are almost perpendicular located to the surface of the crystallites, and tended to regularize, figure 2(d). The observed structuring of the composite were also indicated by increased ultrasonic velocity in the nano-composite sample.

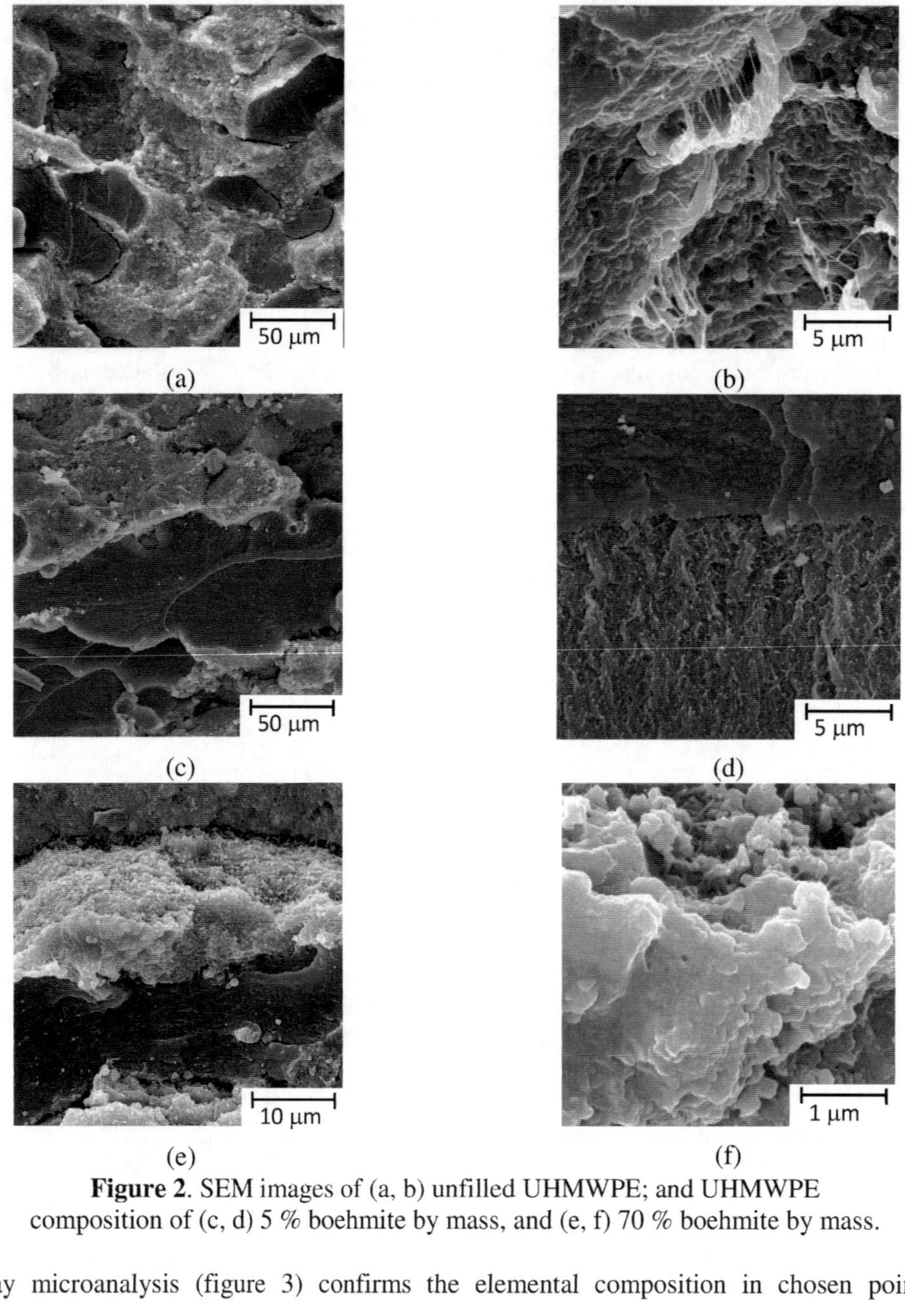

Figure 2. SEM images of (a, b) unfilled UHMWPE; and UHMWPE composition of (c, d) 5 % boehmite by mass, and (e, f) 70 % boehmite by mass.

X-ray microanalysis (figure 3) confirms the elemental composition in chosen points of the morphological polymeric forms.

It is seen, figure 3(a), that for the unfilled UHMWPE, on the various levels of supra-molecular configuration, the spectra are characterized only by presence of carbon, hydrogen and oxygen. In the analogous spectra for supra-molecular structures of UHMWPE composition with the nano-dispersed boehmite, aluminium along with the above elements is well-marked and incorporated into the supra-molecular formations, figure 3(b).

The results of SEM investigation, X-ray diffraction and X-ray spectroscopic analyses show an active influence of the boehmite on the formation of macro- and micro-structures of the polyolefin composite and agree satisfactorily with our research results of a physical and chemical interaction the boehmite and the UHMWPE described before [22].

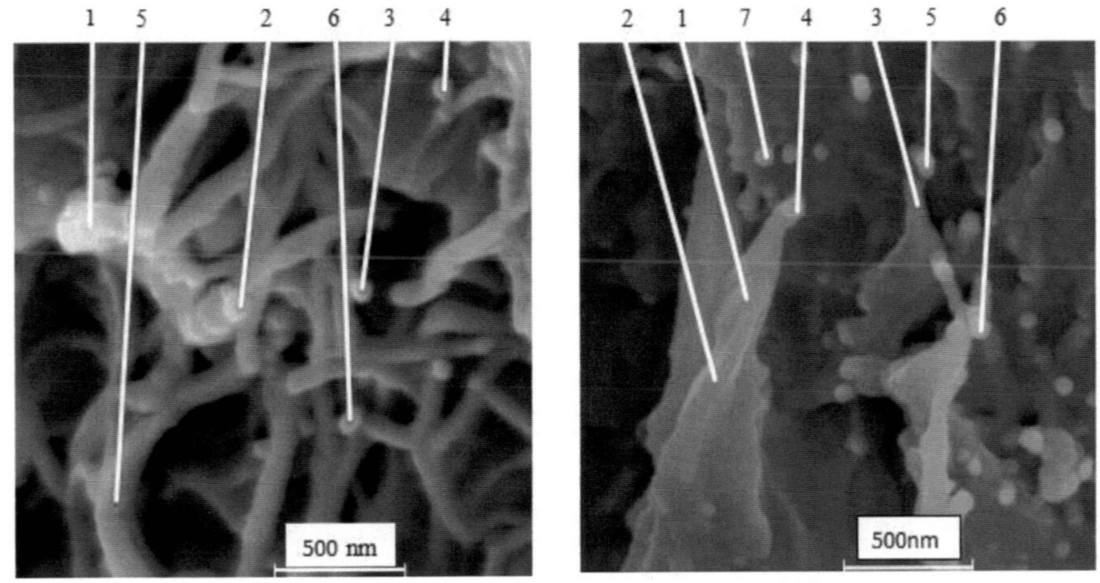

(a) **(b)**

Spectrum	C, O, H % by mass	Al % by mass	Spectrum	C, O, H % by mass	Al % by mass
Spectrum 1	100	0.00	Spectrum 1	rest	6.82
Spectrum 2	100	0.00	Spectrum 2	rest	6.72
Spectrum 3	100	0.00	Spectrum 3	rest	8.43
Spectrum 4	100	0.00	Spectrum 4	rest	6.91
Spectrum 5	100	0.00	Spectrum 5	rest	8.74
Spectrum 6	100	0.00	Spectrum 6	rest	5.02
			Spectrum 7	rest	7.31

Figure 3. SEM micro-images and X-ray microanalysis results of (a) the unfilled UHMWPE, and (b) UHMWPE with 5% boehmite mass.

Concerning the physical, mechanical and deformational properties nano-composites produced, it is necessary to mark the following aspects. For addition of up to 40% by mass of boehmite to the UHMWPE, a reinforcing effect is not observed. The hardness, dynamic elastic modulus and thermal stability of the composites are decreased (table 2, 3). A significant decrease in the temperatures of the phase transitions and the relative deformation increase are observed on the thermos-mechanical curves for the UHMWPE composites with a low boehmite content (figure 4, curve 2). Two regions of α-relaxation, caused by segmental mobility of the chains in a disarranged (amorphous) polymer region, are well marked. The first one is caused by the segmental mobility in the adsorbed (boundary) polymer layer, for which the structure differs from the composition volume. The second point of inflexion corresponds to the characteristic temperature region for the unfilled UHMWPE (figure 4, curve 1). For the addition of ≥ 50 % by mass of nano-ceramic modifier to the polymer a dramatic change of the polymer properties is observed. All the physical and technical properties of the UHMWPE composition with the boehmite are improved (table 2, 3).

The composition of UHMWPE with 70 % by mass of boehmite was tested for resistance to burning under torch flame, according to the GOST 28157-89 Method b standard [21]. The tests showed that the sample placed in the torch flame is carbonized and did not emit smoke. There is no dropping. For the unfilled UHMWPE, there was intensive smoke emission and dropping of melted and burning particles of material was observed.

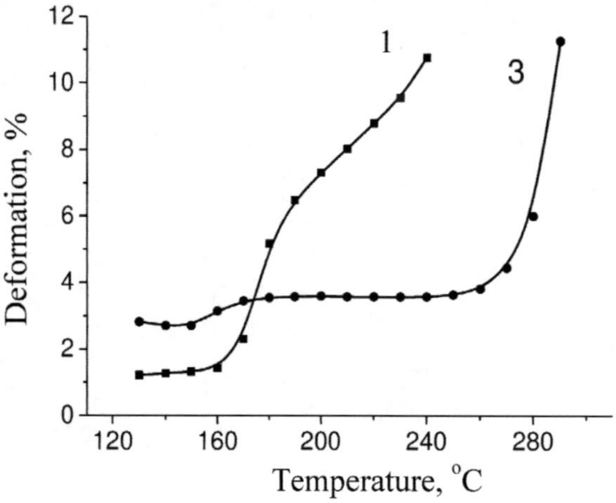

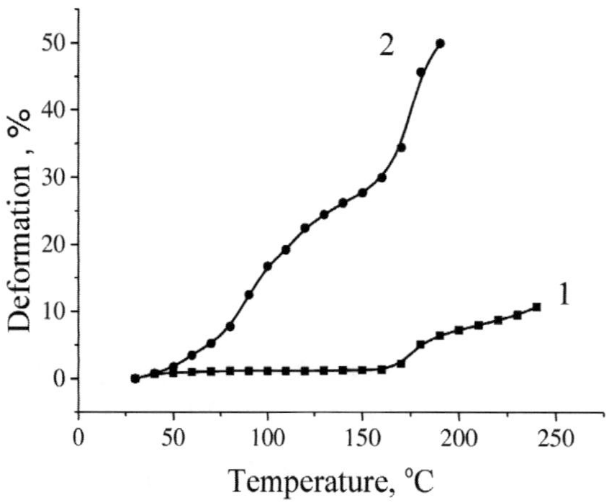

Figure 4. Thermo-mechanical curves of moulded UHMWPE
(1) unfilled, (2) containing 5 % by mass of boehmite and (3) containing 70 % by mass of boehmite.

Table 2. Physical and mechanical properties of composites based on UHMWPE and boehmite.

Boehmite % by mass	Density, kg/m^3	Hardness, MPa	Dynamic elastic modulus, GPa	% Relative deformation at temperature of 220 °C
0.0	1120	145	5.2	8.0
5.0	1130	145	5.1	22.0
30.0	1150	150	4.7	21.0
40.0	1230	160	3.0	20.0
50.0	1380	220	6.3	6.0
70.0	1570	240	6.9	3.0

Table 3. Physical and technical characteristics of composition based on UHMWPE and boehmite.

Boehmite, % by mass	Dielectric relaxation time at frequency of 100 Hz, τ_p, [s·10^{-6}]	Vicat thermal stability, °C
0.0	37.3	230
5.0	9.7	175
30.0	19.5	180
40.0	44.3	190
50.0	65.9	290
70.0	165.8	300

5. Discussion

The experimental results show that for UHMWPE with high volume fractions of boehmite, there is an observed ambiguous change in the properties of the material produced. It may be explained from the perspective of induced interfacial layer. It is known that the structure and properties of polymer composite materials are determined by a number of factors, that is: the structure and properties of its individual components, its mixing conditions and compositions formation, interfacial phenomena, supra-molecular structure and the induced interfacial layer. At the same time, the processes, which take place in the interphase boundary of filler-polymer, are dominant in composite materials properties formation. The fundamentals of the theory of interfacial phenomena in polymer composite materials were laid down since the last century by YuS Lipatov and his school [23], and are still relevant. When the UHMWPE composition is processed with the presence of investigated nano-ceramic filler, some of the macro-molecules, mostly high-molecular fractions, pass into the boundary layers and interact with the polymer [24]. The volume of the polymer is enriched by lower molecular fractions with an increased mobility. It is also known, that the addition of a filler in the rigid chain polymer leads to a reduction of macro-molecule packing density in the boundary layers. As a result, an additional free volume is create, and at sufficiently high volume fractions a looser polymer structure with an increased mobility of the chain components is formed.

The data mentioned above for the thermos-mechanical analysis and the acoustic and electro-physical measurement indicate that for UHMWPE with low volume fractions of boehmite there is still an insufficiently ordered structure. For boehmite concentrations of up to 40 % by mass, a reduction of the dielectric relaxation periods and dynamic elasticity modulus is observed (table 3). The deformation properties of the composites under heating conditions are reduced (table 2, figure 4, curve 2). The physical and mechanical properties and the thermal stability are either reduced or remain at the level of that of the unfilled polymer. The observed properties are not exceptional and accord with the thermoplastic composite materials consisting of a limited amount of excipient and reinforcement, previously developed by the present authors [25, 26]. In general, it is typical for many thermoplastic materials with a low content of dispersed fillers. As for the nano-ceramic modifier boehmite used in this work, along with its many advantages it has a significant disadvantage: the fact that boehmite is not a reinforcing filler this limits its usage in polymeric compositions [27].

In this work it was shown that for UHMWPE with the addition of $\geq$ 50 % by mass of boehmite, when practically all the polymer passes into the interfacial and surface layers, a significant decrease in the mobility of the macro-molecules is observed, as suggested by the dielectric relaxation periods (table 3). The thermos-mechanical analysis data also confirm this hypothesis, see figure 4, curve 3, where the curve trajectory is sharply changed. A significant decrease (of between 1.3 and 2.7 times) of the deformation of the polymeric composite is observed in the whole temperature range studied, and an increase of the thermal stability up to between 290 and 300 °C (table 2, 3) was also observed. Under high volume fraction filling of UHMWPE by boehmite, a completely-bound system is formed, resulting from the physical and chemical transformations in the inter-phase boundary, proceeds with a simultaneous formation of a physical network of boehmite within the composition volume (figure 2e, f). This results in an improvement of all the technical characteristics of the UHMWPE composition. The research results are in satisfactory agreement with the fundamental concepts of the formation of polymer structures in the presence of fillers described in the monograph [28].

6. Conclusions

New highly filled polyolefin composite materials based on UHMWPE and nano-ceramic modifier (boehmite) were developed. It was established that the boehmite actively influences the physical and chemical processes, which take place in the contact region of the nano-ceramic filler – ultrahigh molecular weight polyethylene, under the conditions of composite formation. A correlation of the processes which proceed in interfacial layer, with the structure and properties of the materials produced for various fill fractions of boehmite in the polymer was determined. A sharp increase of the mechanical and thermal properties of the composites is observed for fill fractions of the UHMWPE by the boehmite of equal or greater than 50 % by mass, and reaches a maximum at 70 % boehmite content, when practically the whole polymer is situated on the surface layer in the form of the very thin films. A significant decrease of the macro-molecule mobility in this layer, in conjunction with the formation of a physical network of the active reinforcement within the composition volume, leads to the formation of a completely-bound polymer structure within composition material. As a consequence, the density of the composite of UHMWPE filled with 50 – 70 % by mass of boehmite is increased by 1.2–1.4 times, the dynamic elastic modulus by 1.2–1.3 times, the hardness by 1.5–1.7 times, and the thermal stability by 60–70 °C.

These increased physical, mechanical and thermal characteristics (non-halogenated fire retardant) allow the highly filled composites based on the UHMWPE and the boehmite to be considered for functional designation for workpieces and details of the constructions used in various fields of science, technics and industry. For instance, high requirements for burnability and environmental friendliness are specified for modern materials used in electronic devices constructions and electro-technics. Valid prescriptions, including EU directive of the limiting usage dangerous materials, prohibit the usage of additives with bromine and chlorine content. Taking into account the increased level of fire-resistance of the UHMWPE boehmite composite, this material is able to replace the polymeric compositions used before in mentioned above technical fields.

References

[1] Mayer EA, Dudchenko VK, Poddubnyak AN and Arkatov OL 2003 *Plasticheskie massy* [*Plastics*] **8** 3–4 (in Russian)

[2] Kurtz SM, 2004, *The UHMWPE Handbook: Ultra-High Molecular Weight Polyethylene in Total Joint Replacement*. Academic Press, New York.

[3] Raspopov LN and Belov G 2008 *Plasticheskie massy* [*Plastics*] **5** 13–9 (in Russian)

[4] Aulov VA, Makarov SV, Kuchkina IO, Ozerin AN and Bakeev NF 2000 *Vysokomolekuljarnye soedinenija* [*Macromolecular compounds*] *Series A* **42**(11), 1843–50 (in Russian)

[5] Baronin GS, Stolin AM and Pugachev DV 2009 *Fizika i tehnika vysokih davlenij* [*High Pressure Physics and Engineering*] **19**(2) 146–55 (in Russian)

[6] Selihova VI, Zubov JuA, Sinevich EA, Chvalun SN, Ivancheva NI, Smol'janova OV, Ivanchev SS and Bakeev NF 1992 *Vysokomolekuljarnye soedinenija* [*Macromolecular compounds*] **34**(2) 92–9 (in Russian)

[7] Andreeva IN, Veselovsky EV, Nalivajko EI, Pechenkin AD, Buhgalter VI and Poljakov AV 1982 *Sverhvysokomolekuljarnyj polijetilen vysokoj plotnosti* [*UHMWPE of high density*], Leningrad, Himija Publ (in Russian)

[8] Grozdova GV and Smirnova TN 1987 *Himicheskaja promyshlennost' za rubezhom* [*Chemical industry abroad*] **299**(11) 1–16 (in Russian)

[9] Seljutin GE, Gavrilov JuJu, Voskresenskaja EN, Zaharov VA, Nikitin VE and Polubojarov VA

2010 *Himija v interesah ustojchivogo razvitija* [*Chemistry for sustainable development*] **18**(3), 375–88 (in Russian)

[10] Nikiforov LA, Borisov RV and Okhlopkova AA 2015 *POLIKOMTRIB-2015: Tezisy dokladov mezhdunarodnoj nauchno-tehnicheskoj konferencii* [*POLIKOMTRIB 2015: Abstracts of the International Scientific and Technical Conference*] 23–26 June Gomel 58 (in Russian)

[11] Gogoleva OV 2015 *POLIKOMTRIB-2015: Tezisy dokladov mezhdunarodnoj nauchno-tehnicheskoj konferencii* [*POLIKOMTRIB 2015: Abstracts of the International Scientific and Technical Conference*], 23–26 June Gomel 126. (in Russian)

[12] Okhlopkova AA, Okhlopkova TA, Borisov RV and Golikov AN 2015 *POLIKOMTRIB-2015: Tezisy dokladov mezhdunarodnoj nauchno-tehnicheskoj konferencii* [*POLIKOMTRIB 2015: Abstracts of the International Scientific and Technical Conference*], 23–26 June Gomel 222 (in Russian)

[13] Zoo VS, An J-W, Lim D-Ph and Lim D-S, 2004, Effect of carbon nanotube addition on tribological behavior of UHMWPE, *Tribology Letters*, **16**(4), 305–9

[14] Aderikha VN, Shapovalov VA, Krasnov AP and Pleskachevsey YN 2008 *Trenie i iznos* [*Friction and wear*] **29**(4) 421–7 (in Russian)

[15] Panin SV, Kornienko LA, Nguyen Xuan Thuc, Ivanova LR and Shilko SV 2014 Role of Micro- and Nanofillers in Abrasive Wear of Composites Based on Ultra-High Molecular Weight Polyethylene *Advanced Materials Research* **1040** 148–54

[16] Panin SV, Kornienko LA, Poltaranin MA, Nguen ST, Ivanova LR, Shil'ko SS and Pleskachevskij JuM 2015, *POLIKOMTRIB-2015: Tezisy dokladov mezhdunarodnoj nauchno-tehnicheskoj konferencii* [*POLIKOMTRIB 2015: Abstracts of the International Scientific and Technical Conference*] 23–26 June Gomel 7 (in Russian)

[17] Sudnik LV, Il'iushchenko AF, Sudnik LV, Mazalov IuA and Bersh AV 2010 *Funktsional'nye materialy na osnove gidroksida aliuminiia* [*Functional materials based on aluminum hydroxide*] Belaruskaia navuka publ, Minsk (in Russian)

[18] Stoe WinXPOW. Version 1.04. Stoe& Cie GmbH, Darmstadt (Germany), 1999

[19] Jenkins R, Anderson R ad McCarthy GJ 1999 *Use of the Powder Diffration File JCPDS* International Centre for Diffration Data

[20] GOST 15088-83 1991 *Izdatel'stvo standartov Publ* Moscow (in Russian)

[21] GOST 28157-89 1989 *Izdatel'stvo standartov Publ* Moscow (in Russian)

[22] Dubkov VI, Il'jushhenko AF, Komarevich VG, Krut'ko NP and Sudnik LV 2009 *Materialy. Instrumenty. Tehnologii* [*Materials. Instruments. Technology*] **14**(3) 14–21 (in Russian)

[23] Lipatov YuS 1980 *Mezhfaznye javlenija v polimerah* [*The interphase phenomena in polymers*] Naukova Dumka, Kiev (in Russian)

[24] Fabulyak FG 1983 *Molekuljarnaja podvizhnost' polimerov v poverhnostnyh slojah* [*The molecular mobility in the surface layers of polymers*] Naukova Dumka, Kiev (in Russian)

[25] Dubkov VI, Krutko NP and Minkevich TS 2006 *Materialy 26-oj Mezhdunarodnoj konferencii «Kompozicionnye materialy v promyshlennosti»* [*Proc. 26th Int. Conf. Composite Materials in Industry*], 29 May–2 June, Yalta, Crimea 60–2 (in Russian)

[26] Dubkov VI, Krutko NP, Kukareko VA and Komarevich VG 2009 *Materialy 29-oj Mezhdunarodnoj konferencii «Kompozicionnye materialy v promyshlennosti»* [*Proc. 29th Int. Conf. Composite Materials in Industry*] 1–5 June, Yalta, Crimea, 121–4 (in Russian)

[27] Katz GS and Milewski DV 1981 *Napolniteli dlja polimernyh kompozicionnyh materialov / Spravochnoe posobie* [*Fillers for polymer composites: Handbook*], Himija Publ, Moscow (in Russian)

[28] Lipatov YuS 1977 *Fizicheskaja himija napolnennyh polimerov* [*Physical chemistry of the filled polymers*] Himija Publ, Moscow (in Russian)

Institute of Physics
Dirac House, Temple Back
Bristol BS1 6BE UK

ISSN: 1757-8981
ISBN 978-1-5108-3176-6

CURRAN ASSOCIATES INC.
proceedings.com

Intelligent Transportation Systems 2018

Transportation Research Record: Journal of the Transportation Research Board

Volume 2672, Issue 19

SAGE Publishing